Preface

This solution manual is a companion book written by the authors of "Understanding Physics like a Nerd without Becoming One &More". The character of the book solves the problems that were assigned at the end of each chapter. The authors believe their readers will be inspired by the tactics employed by Cassie to tackle the problems based on the lessons she learned from Professor John.

ISBN: 978-1520502908

Table of Contents

Solutions to Problems and Exercises

Chapter One: Fundamentals of Physics

1. Which object weights more (a) 0.5kg of gold, (b) 0.5kg of grass, (c) 0.5kg of cotton, (d) 0.5kg of iron, or (e) none of the above?

Solution:
Fact #1: all masses are the same,
Fact #2: acceleration due to gravity, g is constant,
Fact #3: weights unknown,
The magnitude of the weight of an object is given by W=mg,
I know the weight of an object is determined by its mass and the acceleration due to gravity. Since all objects have the same mass, the weights of objects are the same. Hence the answer is choice e.

2. Which of the following has the largest volume (a) 0.5kg of gold, (b) 0.5kg of grass, (c) 0.5kg of brass, (d) 0.5kg of iron, (e) none of the above?

Solution:
Fact #1: all masses are the same
Fact #2: density of gold= 19.32 g/cm^3,
Fact #3: density of grass= 0.15 g/cm^3
Fact #4: density of brass=8.4 g/cm^3
Fact #5: density of iron =7.85 g/cm^3.

I understand that volume describes how much space the object occupies. The volume of an object is the ratio of its mass to its density. This means the volume of the object with the lowest density is the largest; therefore, (b) is the correct answer.

3. Which of the following is a base (fundamental) physical quantity (a) volume, (b) speed, (c) temperature or (d) acceleration?

Solution:
Fact #1: volume is length times length times length, which means it can be expressed in another quantity; hence, volume is not a base quantity.
Fact #2: speed is length per unit time; hence, speed is not a base quantity.
Fact #3: temperature does not have anything to describe it other than itself.
Fact #4: acceleration is length per unit time per unit time; hence, acceleration is not a base quantity.

I know a base quantity can only be described by itself. Since temperature cannot be expressed by any other quantity other than itself, temperature is the only base quantity in the above list of choices.

4. Cassie participated in a 15 kilometers race at a Turkey Trot annual event. What was her distance (a) in meters, (b) in miles, and (c) in yards?

Solution:
Fact #1: distance=15km
Fact #2: conversion factor: 1km=1,000m
(a). I understand I must use a conversion factor to change one system of units to another.
Thus, 15km=(15km)(1000m/1.0km)=15,000m
(b). Fact #3: conversion factor: 1mile=1.61km
Thus, 15km=(15km)(1mile/1.61km)=9.31miles
(c) Fact #4: conversion factor: 1km=1,093 yards
Thus, 15km=(15km)(1093yd/1.0km)=16,395 yards

5. Is the following equation dimensionally correct? Explain why or why not.

$$x(t) = \frac{1}{2}at^2 + v_0$$

Solution:
Fact #1: dimension of x(t) is length (L).
Fact #2: dimension of velocity is length per unit time (L/T).
Fact #3: dimension of acceleration is length per unit time per unit time (L/T^2).
Fact #4: dimension of time is T.

Since terms must have the same dimensions to be added; each term in the equation must have the dimension of x(t) which is [L]. Substituting these into the equation above give:

$$[L] = \left[\frac{L}{T}\right] + \left[\frac{L}{T^2}\right][T^2]$$

The dimension of each term on the right side of the equation does not reduce to the same dimension as it is on the left-hand side; hence, the above equation is not dimensionally correct.

6. If the heart rate of an athlete is 60 beats per minute, how many beats does the athlete's heart beat in a year?

Solution:
Fact #1: Heart rate is 60 beats per minute.
Fact #2: Number of days in a year is 365.242 days.
Fact #3: Number of hours in a day is 24 hours.
Fact #4: Number of minutes in one hour=60 minutes.
The central point here is to find how many minutes there are in a year. Once that is known, then I will multiply by the number of beats per minute. Thus, the total number of beats in a year = (365.242 days)(24 hours/day)(60minutes/hour)(60 beats/minute)=31,556,908.8 beats.

7. If the diameter of a carbon nanotube is on the order of nanometers, and the wavelength of low frequency radio waves is about 100 km, how many orders of magnitude larger are radio waves than carbon nanotubes?

Solution:
Fact #1: diameter of carbon nanotube is $\sim 10^{-9}$m
Fact #2: wavelength of low frequency radio waves is 100km=10^5m
To find out how large the radio waves are compared to carbon nanotubes, I simply should obtain the ratio of the wavelength of radio waves to that of the diameter of carbon nanotubes. Thus, the ratio of wavelength of radio waves to the diameter of carbon nanotubes is:

$$\frac{10^5}{10^{-9}} = 10^{14}$$

Hence radio waves are 14 orders of magnitude bigger than carbon nanotubes.

8. If the speed of light in vacuum is 3x10^8 m/s, what is its speed in miles per hour?

Solution:
Fact #1: speed of light is 3x10^8 m/s
Fact #2: 1mile=1,609.344m
Fact #3: one hour=3,600 seconds
This is normally done by using conversion factors that were covered in the first chapter of the book. Thus, the speed of light is

$$(3\text{x}10^8\ {}^{\text{m}}/_{\text{s}})\left(\frac{\text{mi}}{1609.344\text{m}}\right)\left(\frac{3600\text{s}}{\text{h}}\right) = 671{,}080{,}887.62\frac{\text{mi}}{\text{h}}$$

9. If you received a 20% raise on your $50,000.00 annual salary, what will be your new salary?

Solution:
This is a generous raise; I wish I could get that much of a raise every year.
Fact #1: annual salary is $50,000.00
Fact #2: raise is 20%
Fact #3: new salary is unknown

The new salary =$50,000.00+(20/100)($50,000.00)
=$50,000.00+$10,000.00
=$60,000.00.

10. The volume (V) of a certain fluid is changing as a function of time (t) according to the equation

$$V(t) = Ct^3 + B\frac{1}{t^2} + A$$

If t is time in seconds and V is given in m^3, determine the units of A, B and C.

Solution:
Fact #1: The unit for volume is, V =m^3 (given).
Fact #2: The unit for time is, t =s (given).

I know I can answer this problem by using dimensional analysis. I remember that the dimension of every term on the right-hand side of the equation must reduce to the same dimensionality of the left-hand side. Here I have three terms on the right-hand side. The unit of each term must be the same. This means:

i. A must have the units of m^3.
ii. B/t^2 must have the units of m^3
iii. Ct^3 must have the units of m^3

Solving for the units of B and C yields the following results.
B must have the units of m^{3}s^2 and
C must have the units of m^3/s^3.

Chapter Two: Motion along a Straight Line

1. Choose the correct answer: If an object has a non-zero acceleration: (a) It covers the same distance in equal intervals of time, (b) It has a uniform velocity, (c) It covers different distances in equal intervals of time and (d) its final velocity is the same as its initial velocity.

Solution:
Here I have to simply look at the definition of acceleration, which is the rate of change of velocity. If the velocity is changing, then it means the object is travelling different distances in the same interval of time. This means one cannot move the same distance during the same interval of time. Hence the correct answer is (c). Since choices (a), (b) and (d) are describing the same thing, I could have also arrived at the correct answer by the process of elimination.

2. Two families, A and B who live 360 km apart on a straight highway, planned to meet for a retreat somewhere in between. If they start their trips at 2:00 PM with family A driving at 100km/h and family B at 80 km/h; (a) at what time will they meet, and (b) where will they meet?

Solution:
(a) I understand that both families must share the 360km distance between them. My intuition tells me that the quickest way to answer this question is to assume a super vehicle that has the same speed as the sum of the two speeds, which gives 180km/h. So the time in which the two families meet is found by dividing the distance by the super speed, i.e.,

$$t = \frac{360\text{km}}{180\,{}^{\text{km}}/_{\text{h}}} = 2\text{h}$$

However, I have learned in previous discussions that my intuition does not always give the correct result. Thus, let me solve this in the non-lazy way.
Fact #1: distance to travel is 360 km,
Fact #2: speed of family A; v_1=100 km/h,
Fact #3: speed of family B; v_2=80 km/h,
Suppose family A travels x kilometers when they meet the second family; then family B will travel 360-x kilometers. Here I can write:

$$\text{x} = \text{v}_1\text{t} \qquad \text{and} \qquad 360 - \text{x} = \text{v}_2\text{t}$$

Where t is the time it took the two families to meet. By combining these two equations, I can solve for the time t to obtain

$$\text{t} = \frac{360\text{km}}{\text{v}_1 + \text{v}_2} = \frac{360\text{km}}{100\,{}^{\text{km}}/_{\text{h}} + 80\,{}^{\text{km}}/_{\text{h}}} = 2\text{h}$$

This was the same answer that I obtained by my super vehicle assumption. It seems that intuition can be correct sometimes. This means the families will meet two hours after they leave their homes. Hence they will meet at 4:00 PM.

(b) In two hours, family A will travel (100km/h)(2h)=200km. Thus the families meet at 200km from the home of family A.

3. Two families are headed for a retreat 300 km away. One of the families got a head start and began driving at 1:30PM with a speed of 90 km/h. The second family left 30 minutes after and started driving at 120 km/h. (a) How long does it take the second family to catch up with the first family, (b) where will the second family catch the first family, and (c) if they continue traveling nonstop towards their destination at their respected speeds, at what time will each family reach their destination?

Solution:
Fact #1: speed of the first family is v_1=90 km/h,
Fact #2: speed of the second family is v_2=120 km/h,
Fact #3: distance travelled by the first family is x_1= unknown,
Fact #4: distance travelled by the second family is x_2=unknown

Fact #5: time of travel of second family is t= unknown

(a) In this case, the second family must cover whatever the first family covered in the first 30 minutes plus whatever distance they travelled once both families are on the road. Unlike problem 3, here each family is travelling in the same direction. When the second family begins driving, the first family has been driving for one-half of an hour. During the first 30 minutes, the distance travelled by the first family is obtained by

$$d = v_1 t = \left(90\frac{km}{h}\right)(0.5h) = 45km$$

Now instead of the super-fast car in the previous problem, I can think of a super slow car that is driving at a speed that is the difference of the two speeds, i.e.

$$v = 120\,{}^{km}/_{h} - 90\,{}^{km}/_{h} = 30\,{}^{km}/_{h}$$

At this speed, the time it takes to travel the 45km by the super slow car will be

$$t = \frac{d}{v} = \frac{45km}{30\,{}^{km}/_{h}} = 1.5h$$

The late staring family will catch up with the other family in one hour and thirty minutes. However, I want to confirm this by doing the alternate approach. The distance traveled by each family is obtained by using one of the equations of motion, i.e.

$$x_1 = v_1 t \qquad \text{and} \qquad x_2 = v_2 t$$

Here x_2 and x_1 are related by

$$x_2 = x_1 + d$$

Substituting for x_1 and x_2 from the above relationship and solving for time yields

$$t = \frac{d}{v_2 - v_1} = \frac{45km}{120\,{}^{km}/_{h} - 90\,{}^{km}/_{h}} = 1.5h$$

This means it will take the late starting family an hour and a half to catch up with the early bird family which, was the same result I obtained with the super slow car approach. It looks that my intuition on this one is also on the mark. I have two for two rights, but I have learned not to brag from Professor John so I am not doing that.

(b) For the second part, I just have to find x_2 by using the above equation, i.e.

$$x_2 = v_2 t = (120\,{}^{km}/_{h})(1.5h) = 180km$$

(c) Here I have to find how long it takes each family to travel 300km. If I assign t_1 and t_2 to be the times of travel of each family respectively, then I can write

$$t_1 = \frac{300km}{90\,{}^{km}/_{h}} = 3h \text{ and } 20 \text{ minutes}$$

$$t_2 = \frac{300\text{km}}{120\,{}^{\text{km}}/_{\text{h}}} = 2\text{h and 30 minutes}$$

Thus, the first family will get to their destination at 4:50PM, and the second family at 4:30 PM.

4. An object that has been dropped from a tall building was observed to pass though a window that is 10m high. The observer inside records the time it took to pass by the window to be 0.20s with his smart phone. (a) How fast was the object falling when it appeared at the top of the window, (b) how fast was the object falling when it disappeared at the bottom of the window, (c) from what height above the top edge of the window was the object dropped, and (d) if the object was heard to hit the ground 3.0s after it disappeared, what is the total height of the building? Hint: assume free fall motion with g=10m/s² downwards.

Solution:
Fact #1: acceleration due to gravity is, $g= -10\ m/s^2$,
Fact #2: time is, $t= 0.20s$,
Fact #3: displacement is, $\Delta h=-10m$,
Fact #4: initial velocity at the top of the window is v_1=unknown,
Fact #5: final velocity at the bottom of the window is v_2=unknown.

Note: I will use the sign convention that downward motion is negative. Since there are two unknowns, I need to come up with an additional equation. From the information given, I can obtain the average velocity by dividing the displacement by the duration, i.e.

$$\bar{v} = \frac{\Delta h}{t} = \frac{-10m}{t} = -50\,{}^{\text{m}}/_{\text{s}}$$

Because the acceleration is constant, I can also obtain the average velocity of v_1 and v_2 by using:

$$\bar{v} = \frac{v_2 + v_1}{2}$$

Applying the value obtained above I can write the average velocity as

$$\bar{v} = \frac{v_2 + v_1}{2} = -50\,{}^{\text{m}}/_{\text{s}} \qquad (1)$$

This equation will be the first equation that relates v_1 and v_2. The second equation is obtained by applying the definition of acceleration, i.e.

$$a = \frac{v_2 - v_1}{t} = -10\,{}^{\text{m}}/_{\text{s}^2} \qquad (2)$$

Simplifying the last two equations and substituting for t, I can write

$$v_2 + v_1 = -100\,{}^{\text{m}}/_{\text{s}}$$

$$v_2 - v_1 = -2.0\,{}^{\text{m}}/_{\text{s}}$$

Solving for v_1 and v_2 yields

$$v_2 = -51.0\ {}^{\mathrm{m}}/_{\mathrm{s}} \qquad \text{and} \qquad v_1 = -49.0\ {}^{\mathrm{m}}/_{\mathrm{s}}$$

The negative signs indicate downward motion. In order to find the height above window, I need to write the following facts.
Fact #6: initial velocity v_0 is 0, because the object was just dropped,
Fact #7: acceleration is still the same at -10m/s^2,
Fact #8: v_1 serves as the new final velocity,
Fact #9: height above the window is Δh_2=unknown.
The equation of motion that connects facts 6 through 9 is the following,

$$v^2 = v_0^2 + 2a\Delta x$$

Rearranging and substituting the facts, yield

$$\Delta h_2 = \frac{v_1^2 - v_0^2}{2a} = \frac{(-49\ {}^{\mathrm{m}}/_{\mathrm{s}})^2}{2\left(-10\ {}^{\mathrm{m}}/_{\mathrm{s}^2}\right)} = -120.05\mathrm{m}$$

Thus, the object has fallen from a height of 120.05 m above the top of the window.

Now let me consider the motion of the object after it has disappeared from bottom of the window. For this part of the motion, v_2 serves as an initial velocity. The following equation of motion has all of the parameters except the one that I am trying to find.

$$\Delta x = v_0 t + \frac{1}{2}at^2$$

Substitution all the known quantities yield,

$$\Delta h_3 = (-51\ {}^{\mathrm{m}}/_{\mathrm{s}})(3\mathrm{s}) + \frac{1}{2}\left(-10\ {}^{\mathrm{m}}/_{\mathrm{s}^2}\right)(3\mathrm{s})^2 = -198\mathrm{m}$$

Thus, the total height of the building is 120.05m+10m+198m=328.05m

5. In a relay race of four runners; Liz, Betty, Mary and Amelia ran at 15 km/h, 20 km/h, 25km/h and 30 km/h respectively. If everyone runs the same distance during the race, (a) what is the magnitude of the average velocity (speed) of this relay team, and (b) why can't you simply add the four speeds and obtain the average velocity of the relay team?

Solution:
Fact #1: Liz's speed is v_1=15km/h,
Fact #2: Betty's speed is v_2=20km/h,
Fact #3: Mary's speed is v_3=25km/h,
Fact #4: Amelia's speed is v_4=30km/h.
Fact #5: distance of relay is unknown,
Fact #6: time of each runner is unknown.

(a) The point to remember is that each runner is covering the same distance in different times. The average velocity is the total distance divided by the total time, i.e.

$$\bar{v} = \frac{\Delta x}{\Delta t}$$

Where Δx and Δt are the total distance and time respectively. If I represent the distance that each racer runs by *d*, then Δx=4*d*, and each respective time will be the runners distance divided their speed so that;

$$\boldsymbol{\Delta t} = \frac{d}{v_1} + \frac{d}{v_2} + \frac{d}{v_3} + \frac{d}{v_4}$$

Hence

$$\bar{v} = \frac{4d}{\frac{d}{v_1} + \frac{d}{v_2} + \frac{d}{v_3} + \frac{d}{v_4}} = \frac{4}{\frac{1}{v_1} + \frac{1}{v_1} + \frac{1}{v_1} + \frac{1}{v_1}} = 21.05\,{}^{\text{km}}/_{\text{h}}$$

(b) Taking the sum of the different speeds and dividing by 4 gives 22.5 km/h. This approach cannot give the correct result because each runner does not complete the race in same amount of time.

6. Savannah was driving her new car and stopped at a traffic light. When the light turned green, an approaching truck passed her with a velocity of 72km/h. If Savannah accelerates at a constant rate and overtakes the truck 200m down the road. (a) What was her acceleration, and (b) how fast was she travelling at the time she overtook the truck?

Solution:
Fact #1: initial velocity of Savannah is v_0=0,
Fact #2: velocity of truck is v_T=72 km/h=20m/s,
Fact #3: distance travelled by both is 200m,
Fact #4: acceleration of Savannah is unknown,
Fact #5: time unknown,
Fact #6: final velocity of Savannah is unknown.

In the beginning the truck will be ahead; however, because Savannah is accelerating, she will eventually catch up with the truck. Since I know the truck is travelling at constant velocity, I can find the time by dividing the distance by the speed, i.e.

$$t = \frac{200\text{m}}{20\,{}^{\text{m}}/_{\text{s}}} = 10\text{s}$$

Since the distance, the initial velocity and the time are all known, the correct equation of motion to use is:

$$\Delta x = \text{v}_0\text{t} + \frac{1}{2}\text{at}^2$$

Now I can solve for the acceleration by substitution, i.e.

$$a = \frac{2\Delta x}{t^2} = \frac{2(200\text{m})}{(10\text{s})^2} = 4\,{}^{\text{m}}/_{\text{s}^2}$$

The final velocity of Savannah is obtained by simply using the definition of acceleration, which is the rate of change of velocity. Rearranging, therefore, yields

$$v = v_0 + at = 0 + \left(4\,{}^{\text{m}}/_{\text{s}^2}\right)(10\text{s}) = 40\,{}^{\text{m}}/_{\text{s}}$$

7. A block sliding on a horizontal surface has an initial speed of 10.0 m/s. The block travels a distance of 25.0m as it slows to a stop. What distance would the block have travelled if its initial speed had been 20.0 m/s?

Solution:
Fact #1: initial velocity is v_0=10.0m/s,
Fact #2: final velocity is v=0,
Fact #3: stopping distance is 25.0m,
Fact #4: in the second situation, initial velocity is v_0'=20.0m/s,
Fact #5: in the second situation, final velocity is v'=0,
Fact #6: stopping distance is unknown.

Here I have to make a note that the block is decelerating at a constant rate. From the information given, I can obtain the acceleration of the block by using the following equation of motion.

$$v^2 = v_0^2 + 2a\Delta x$$

Solving for acceleration gives

$$a = \frac{v^2 - v_0^2}{2\Delta x} = \frac{0 - (10\,{}^{\text{m}}/_{\text{s}})^2}{2(25\text{m})} = -2\,{}^{\text{m}}/_{\text{s}^2}$$

The acceleration of the block remains the same. The newly found acceleration becomes a fact of the problem, i.e.
Fact #7: acceleration is $a=-2\text{m/s}^2$,
Fact #8: the new distance is $\Delta x'$=unknown.
I can still use the above equation of motion to find the new displacement, i.e.

$$\Delta x' = \frac{v^2 - v_0^2}{2a} = \frac{0 - (-20\,{}^{\text{m}}/_{\text{s}})^2}{2(-2\,{}^{\text{m}}/_{\text{s}^2})} = 100\text{m}$$

Thus, the block will cover 100m when it stops. I see that even though the speed was double, the stopping distance was quadrupled.

8. Kevin was driving on a highway at 108 km/h when he saw an accident ahead of him. He immediately applied the brake and decelerated at a constant rate. If he stops in a distance of 150m (a) how long did it take him to stop, and (b) what was his acceleration?

Solution:
This is similar to Problem #7.
Fact #1: initial velocity is v_0=108km/h=30m/s,
Fact #2: final velocity is v=0,
Fact #3: Stopping distance is Δx=150m,
Fact #4: time is unknown,

Fact #5: acceleration is unknown.

I think it is better to solve the second question before the first one. Solving for the acceleration I obtain,

$$a = \frac{v^2 - v_0^2}{2\Delta x} = \frac{0 - (30\,{}^{\mathrm{m}}/_{\mathrm{s}})^2}{2(150\mathrm{m})} = -3\,{}^{\mathrm{m}}/_{\mathrm{s}^2}$$

Now to find time, I only need to use the newly found acceleration as a fact. Defining acceleration as the rate of change of velocity, I can write,

$$a = \frac{v - v_0}{t}$$

Solving for t yields,

$$t = \frac{v - v_0}{a} = \frac{0 - 30\,{}^{\mathrm{m}}/_{\mathrm{s}}}{-3\,{}^{\mathrm{m}}/_{\mathrm{s}^2}} = 10\mathrm{s}$$

9. When Sandi, Rhonda and Jerry were driving three different cars on a straight highway at 120km/h within 60m of each other, a situation was created that required all of them stop without hitting each other. If three of them have different reaction times (the time necessary to move the foot from the accelerator to the brake), how should the reaction times of Sandi and Jerry compare so that Rhonda could avoid a collision with either Sandi or Jerry. Jerry is leading, Rhonda is in the middle and Sandi is last.

Solution:
Fact #1: all are driving in close proximity.
Fact #2: the desired outcome for Rhonda is to avoid hitting Jerry and to escape being hit by Sandi.
I say that every driver must remain calm and be fully alert. Rhonda has to watch the actions of both Sandi and Jerry and react accordingly. For Rhonda to avoid hitting Jerry, Jerry should have a long reaction time so he continues to get farther from Rhonda. In order to avoid hitting Rhonda, Sandi must stop quickly, which means her reaction time should be very fast. So Rhonda will be ok, if Jerry has a slow reaction time and Sandi has a fast reaction time.

10. A tennis ball that has been dropped from 20.0m above the ground stays in contact with the ground for 0.05s before it rebounds to a height of 10.0m. What is the acceleration of the ball during its contact with the ground?

Solution:
Fact #1: first distance is Δx=-20m,
Fact #2: acceleration is $g=-10m/s^2$,
Fact #3: initial velocity is 0,
Fact# 4: final velocity is v=unknown,
Fact #5: time is t=0.05s,
Fact #6: second distance is Δx=10m,

This is a case of free fall where the object is subject only to the force of gravity. In order to find the acceleration, I have to obtain the velocity when the ball hits the ground and when it leaves the ground. The right equation of motion to use to obtain both these velocities is

$$v^2 = v_0^2 + 2a\Delta x$$

Substituting the values and solving for v yields,

$$v = \sqrt{2\left(-10\,{}^{\text{m}}/_{\text{s}^2}\right)(-20\text{m})} = 20\,{}^{\text{m}}/_{\text{s}}$$

The direction of v is downwards. Let v_0' be the velocity with which the ball left the ground when it rebounds, where v_0' is unknown. Let also v' be the velocity when the ball rebounds to 10 m, where v'=0. Substituting these values into the equation below and solving for v_0' results

$$v' = \sqrt{v'^2_0 + 2a\Delta x}$$

$$0 = \sqrt{v'^2_0 + 2(-10\,{}^{m}/_{s^2})(10\text{m})}$$

Hence

$$v_0' = \sqrt{200\,{}^{\text{m}^2}/_{\text{s}^2}} = 14.14\,{}^{\text{m}}/_{\text{s}}$$

The direction of v' is upwards.
Thus acceleration is:

$$a = \frac{v'_0 - v}{t} = \frac{14.14\,{}^{\text{m}}/_{\text{s}} - (-20\,{}^{\text{m}}/_{\text{s}})}{0.05\text{s}} = 682.8\,{}^{\text{m}}/_{\text{s}^2}$$

11. John is an elderly gentleman and one sunny day he was driving at 108 km/h on the highway when he saw an accident ahead of him and wanted to stop as soon as possible. If his reaction time was 0.5s, and he decelerated at $5m/s^2$ once he stepped on the brake, what is his total stopping distance?

Solution:
Fact #1: initial velocity is 108km/h=30m/s,
Fact #2: reaction time is t_r=0.5s,
Fact #3:acceleration is a=-5m/s^2,
Fact #4: total distance is unknown,
Fact #5: final velocity is 0.

Here I have to recognize that there two types of motion. One before the brake is applied and the other after the brake has been applied. The motion before the brake is applied is one of constant velocity motion. The total distance is the sum of the distances travelled before and after the brake has been applied. Let x_1 and x_2 be distances covered before and after the brake has been applied respectively. Before the brake was applied, the velocity was constant (30m/s). Thus, I get

$$x_1 = (30\,{}^{m}/_{s})(0.5s) = 15m$$

Once the brake is applied, the motion is one of deceleration, i.e.

$$x_2 = \frac{v^2 - v_0^2}{2a} = \frac{0 - (30\,{}^{m}/_{s})^2}{2(-5\,{}^{m}/_{s^2})} = 90m$$

Thus, the total distance is:

$$x = x_1 + x_2 = 15m + 90m = 105m$$

Additional Tidbits

(a) What physical quantity does the area under the curve of acceleration vs time graph represent?
(b) When an object is moving on a straight line and covers different distances in the same interval of time, how do you describe the motion of the object?
(c) If the slope of velocity versus time graph of a moving object is zero, what does it tell you about the motion of an object?
(d) What do you call the magnitude of displacement?
(e) What do you call the magnitude of velocity?

Answers to Additional Tidbits

(a) The area under the curve of acceleration versus time graph represents velocity.
(b) The object has non-zero acceleration.
(c) The object is moving with uniform velocity.
(d) Distance is the magnitude of displacement.
(e) Speed is the magnitude of velocity.

Chapter 3: Vectors

1. Is a displacement of 3 blocks east followed by 4 blocks north equal to 7 blocks northeast? Explain why.

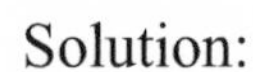

Solution:
Fact #1: first displacement is 3 blocks East,
Fact #2: second displacement is 4 blocks North,
Fact #3: The sum of total displacement is unknown.

I know displacement is a vector quantity, and it must not be dealt with like a scalar quantity. The sum of the two displacements must take into consideration the direction of each displacement. One approach to find the total displacement is to use the graphical method. Let one-half inch represent one block. Then the first displacements can be represented by 1.5 inches east and the second by 2.0 inches north as shown below. The sum of the two displacements is obtained by connecting the tail of the first displacement to the head of the second displacement. Measuring the total displacement of this vector gives 2.5 inches. This means the magnitude of the total displacement was 5 blocks.

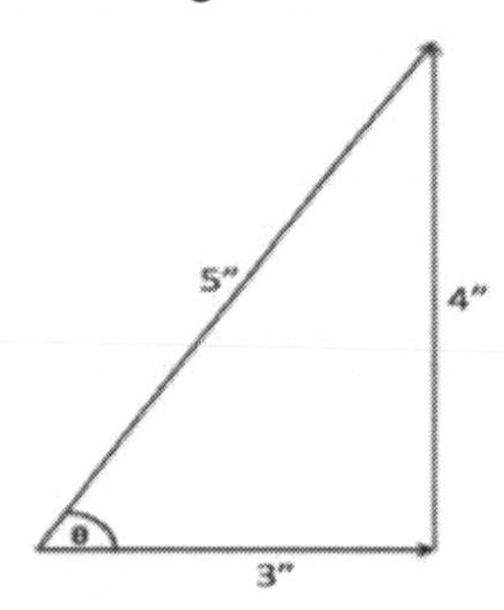

I have now found the displacement. Oh…wait…displacement is a vector quantity and it must have two pieces of information, which means I need to find the angle also. Taking the ratio of opposite to adjacent gives the tangent of the angle and then obtaining the inverse tangent gives the value of the angle, i.e.

$$\tan\theta = \frac{2}{1.5} = {}^{4}/_{3}$$

Hence

$$\theta = \tan^{-1}(4/3) = 53^{0}$$

2. How many possible ways are there to change a vector quantity?

Solution:
Since a vector quantity has magnitude and direction, it can be changed by varying its magnitude alone or by changing its direction or by varying both simultaneously. Therefore there are three possible ways to change a vector quantity.

3. For a given vector,

$$\vec{A} = 8\hat{i} - 6\hat{j}$$

(a) Find the magnitude of A and the unit vector directed along A
(b) Find the magnitude of the unit vector
(c) Find the direction of vector A.

Solution:
a. The magnitude of a vector is obtained by taking the square root of the sum of the squares of the components.
Fact #1: The x-component is 8
Fact #2: The y-component is -6
Thus the magnitude of the vector A is:

$$|\vec{\mathbf{A}}| = \sqrt{(8)^2 + (-6)^2} = 10$$

The unit vector is obtained by dividing the vector by its magnitude. Thus if e_A is the unit vector along A then;

$$\vec{e}_A = \frac{\vec{\mathbf{A}}}{|\vec{\mathbf{A}}|} = \frac{\mathbf{8\vec{i} - 6\vec{j}}}{10} = 0.8\vec{\mathbf{i}} - \mathbf{0.6\vec{j}}$$

b. The magnitude of the unit vector is obtained like any other vector, i.e.

$$|\vec{e_A}| = \sqrt{(0.8)^2 + (-0.6)^2} = 1.0$$

c. The direction of vector A is obtained by taking the inverse tangent of the ratio of the y-component to that of the x-component, i.e.

$$\theta = \tan^{-1}\left({}^{-6}/_{8}\right) = -37^0$$

The vector components indicate that the vector is in the fourth quadrant, this means that the vector lies -37^0 from the positive x-axis.

4. If two vectors of magnitude 30 units and 15 units respectively are added together. (a) What is the possible maximum value of the magnitude of their sum and what is the condition for the maximum value to occur, and (b) what is the possible minimum value of the magnitude of their sum and what is the condition for the minimum to occur?

Solution:
Fact #1: magnitude of first vector is 30,
Fact #2: magnitude of second vector is 15,
I understand that vector addition must take the magnitude and direction of the vectors into account. The sum of any two vectors has maximum magnitude when the two vectors have the same direction. Thus, the possible maximum is 30+15=45.

The sum of two vectors has minimum magnitude when the two vectors are in opposite direction. Thus, the possible minimum magnitude is 30-15=15.

5. A top golfer on the green is having a bad day, and he took three successive strokes to put the ball in the hole. If the first stroke was 10m north, and the second stroke was 20m east and the final stroke was 10 m northeast, what was the magnitude of the displacement of the ball?

Solution:
This is another vector addition, which can be done graphically or analytically.
Fact #1: first displacement is d_1=10m north,
Fact #2: second displacement is d_2=20m east,
Fact #3: third displacement is d_3=10m northeast.
When more than two vectors are added, the analytical method is more convenient than the graphical approach.

vector	east	north
d_1	0	10m
d_2	20m	0
d_3	10mcos45	10msin45

Here, I add the east and north components separately. If d_x and d_y represent the sum of the components directed east and north respectively, then I can write:

$$d_x = 20\text{m} + 10\text{mcos}45 = 27.07\text{m}$$

$$d_y = 10\text{m} + 10\text{msin}45 = 17.07\text{m}$$

Thus, the magnitude of the total displacement is:

$$d = \sqrt{(27.07\text{m})^2 + (17.07)^2} = 32.0\text{m}$$

6. Show that the magnitude of the resultant of two vectors A and B can be given by

$$|\vec{R}| = \sqrt{A^2 + B^2 + 2AB\cos(\theta)}$$

where θ is the angle between the two vectors A and B.

Solution:
Since this is vector addition I can use the analytical approach to do the addition and once I find the sum, then I can obtain the magnitude of the resultant vector.
Fact #1: given two vectors A and B,
Fact #2: the angle between the two vectors is θ
As I have learned, having a visual image of the problem is very helpful to solving the problem. Hence let the two vectors be arranged as shown below.

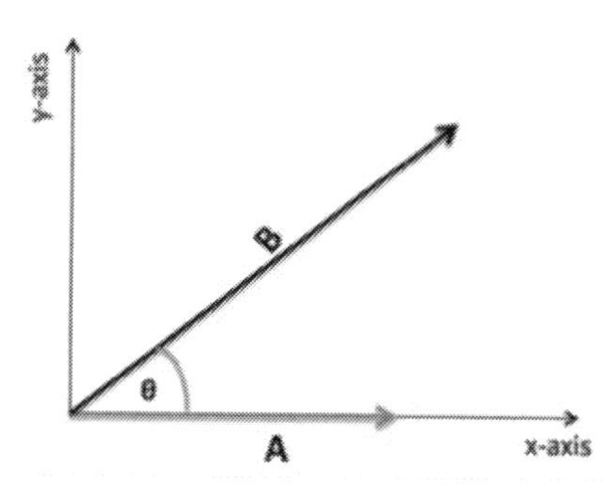

Vector A is made to lie on the x-axis for the sake of convenience. To find the resultant vector, let me break each vector into its components.

Vector	x-component	y-component
A	A	0
B	Bcosθ	Bsinθ

Let R_x and R_y be the sum of the x and y-components respectively, i.e.

$$R_x = A + B\cos\theta$$

$$R_y = B\sin\theta$$

The resultant vector can then be written as

$$\vec{R} = (A + B\cos\theta)\vec{\imath} + B\sin\theta\,\vec{\jmath}$$

The magnitude of the resultant is:

$$|\vec{R}|^2 = (A + B\cos\theta)^2 + (B\sin\theta)^2 = A^2 + 2AB\cos\theta + B^2\cos^2\theta + B^2\sin^2\theta$$

By using the identity

$$\cos^2\theta + \sin^2\theta = 1$$

and taking the square root yields the expected result, i.e

$$|\vec{\mathbf{R}}| = \sqrt{A^2 + B^2 + 2AB\cos\theta}$$

7. Find the angle between the following two vectors.

$$\vec{A} = 5\hat{i} + 6\hat{j} + \sqrt{2}\hat{k}$$
$$\vec{B} = -5\hat{i} + 7\hat{j} - \sqrt{2}\hat{k}$$

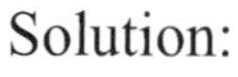

Solution:

I can obtain the angle between the two vectors by using the definition of the dot product, i.e.

$$\vec{A} \cdot \vec{B} = AB\cos\theta$$

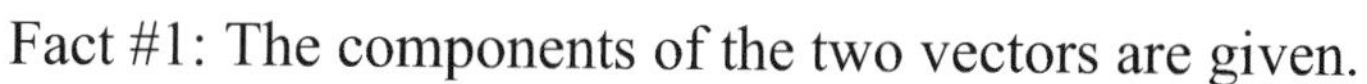

Fact #1: The components of the two vectors are given.

From the above definition, the dot product of two vectors can be written as follows.

$$\vec{A}\cdot\vec{B} = A_xB_x + A_yB_y + A_zB_z$$

Thus

$$\vec{A}\cdot\vec{B} = (5)(-8) + (6)(7) + (\sqrt{2})(-\sqrt{2}) = 0$$

Now setting

$$AB\cos\theta = A_xB_x + A_yB_y + A_zB_z$$

Solving for cosθ yields:

$$\cos\theta = \frac{A_xB_x + A_yB_y + A_zB_z}{|\vec{A}||\vec{B}|} = \frac{0}{|\vec{A}||\vec{B}|} = 0$$

The angle is then obtained by taking the inverse cosine of θ, i.e.

$$\theta = \cos^{-1} 0 = 90^0$$

8. Find the magnitude of the cross (vector) product of the following two vectors.

$$\vec{P} = 4\hat{i} + 4\hat{j}$$
$$\vec{Q} = 5\hat{i} - 5\hat{j}$$

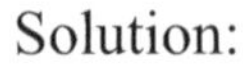

Solution:

By definition the magnitude of the vector product is given by

$$|\vec{A}\times\vec{B}| = |\vec{\mathbf{A}}||\vec{\mathbf{B}}|\sin\theta$$

Since the angle between the two vectors is not given, I have to either find the angle by using the dot product as in Problem 7 above or use the method of determinants to find the cross product. In order to do that, I have to make a slight modification to the given vectors and write

$$\vec{P} = 4\hat{i} + 4\hat{j}$$
$$\vec{Q} = 5\hat{i} - 5\hat{j}$$

Hence

$$\vec{\mathbf{P}}\times\vec{Q} = \begin{vmatrix} \vec{i} & \vec{j} & \vec{k} \\ 4 & 4 & 0 \\ 5 & -5 & 0 \end{vmatrix}$$

$$\vec{i}[(4)(0) - (0)(-5)] - \vec{j}[(0)(5) - (4)(0)] + \vec{k}[(4)(-5) - (4)(5)]$$

$$\vec{\mathbf{P}}\times\vec{Q} = 0\vec{\mathbf{i}} + 0\vec{\mathbf{j}} - 40\vec{\mathbf{k}}$$

Thus, the magnitude of the cross product is 40 $(\text{units})^2$.

9. Show that the vector product of the vectors given in problem 8 is always perpendicular to both vectors.

Solution:
Here I have to show that the dot product of each vector with the cross product is zero.
Fact #1: The components of the two vectors have been given. The cross product has been obtained, i.e.

$$\vec{\mathbf{P}} = 4\vec{\mathbf{i}} + 4\vec{\mathbf{j}} + \mathbf{0}\vec{\mathbf{k}}$$

$$\vec{\mathbf{Q}} = 5\vec{\mathbf{i}} - 5\vec{\mathbf{j}} + \mathbf{0}\vec{\mathbf{k}}$$

$$\vec{\mathbf{P}} \times \vec{\mathrm{Q}} = 0\vec{\mathbf{i}} + 0\vec{\mathbf{j}} - 40\vec{\mathbf{k}}$$

Here I have to show that both the dot and cross product with each vector is zero.

$$\vec{\mathbf{P}} \cdot \left(\vec{\mathbf{P}} \times \vec{\mathrm{Q}}\right) = \left(4\vec{\mathbf{i}} + 4\vec{\mathbf{j}} + \mathbf{0}\vec{\mathbf{k}}\right) \cdot \left(0\vec{\mathbf{i}} + 0\vec{\mathbf{j}} - 40\vec{\mathbf{k}}\right) = [(\mathbf{4})(\mathbf{0})] + (\mathbf{4})(\mathbf{0}) + (\mathbf{0})(\mathbf{-40}) = \mathbf{0}$$

Similarly

$$\vec{\mathbf{Q}} \cdot \left(\vec{\mathbf{P}} \times \vec{\mathrm{Q}}\right) = \left(5\vec{\mathbf{i}} - 5\vec{\mathbf{j}} + \mathbf{0}\vec{\mathbf{k}}\right) \cdot \left(0\vec{\mathbf{i}} + 0\vec{\mathbf{j}} - 40\vec{\mathbf{k}}\right) = [(\mathbf{5})(\mathbf{0})] + (\mathbf{-5})(\mathbf{0}) + (\mathbf{0})(\mathbf{-40}) = \mathbf{0}$$

Thus, the vector product of the two vectors is perpendicular to each vector.

10. Given three vectors A, B and C where

$$\vec{\mathbf{A}} = 4\vec{\mathbf{i}} - 5\vec{\mathbf{j}} + 6\vec{\mathbf{k}}$$
$$\vec{\mathbf{B}} = -6\vec{\mathbf{i}} + \vec{\mathbf{j}} - \mathbf{10}\vec{\mathbf{k}}$$
$$\vec{\mathbf{C}} = k_1\vec{\mathbf{i}} + k_2\vec{\mathbf{j}} + k_3\vec{\mathbf{k}}$$
$$\vec{\mathbf{D}} = k\vec{\mathbf{i}} + k\vec{\mathbf{j}} + k\vec{\mathbf{k}}$$

(a) Find the value of $\vec{\mathbf{C}}$ such that

$$\vec{\mathbf{A}} + \vec{\mathbf{B}} + \vec{\mathbf{C}} = 0$$

(b) Find the value of $\vec{\mathbf{D}}$ such that

$$\vec{\mathbf{A}} \cdot (\vec{\mathbf{B}} + \vec{\mathbf{D}}) = 0$$

Solution:
(a) For the first question, I have to add the vectors as indicated and set each component to zero.
Fact #1: Three vectors and their components are given.
Adding the three vectors yields,

$$\vec{\mathbf{A}} + \vec{\mathbf{B}} + \vec{\mathbf{C}} = \vec{\mathrm{i}}(4 - 6 + \mathrm{k}_1) + \vec{\mathrm{j}}(-5 + 1 + \mathrm{k}_2) + \vec{\mathrm{k}}(6 - 10 + \mathrm{k}_3) = \vec{0}$$

This means the sum of each component must be zero, i.e.

$$4 - 6 + \mathrm{k}_1 = 0$$

$$-5 + 1 + \mathrm{k}_2 = 0$$

$$6 - 10 + k_3 = 0$$

This gives the components of vector C as $k_1 = 2$, $k_2 = 4$ and $k_3 = 4$ so that vector C is given by

$$\vec{C} = 2\vec{i} + 4\vec{j} + 4\vec{k}$$

(b) For the second part I have to find the dot product first.

$$\vec{A} \cdot (\vec{B} + \vec{D}) = (4\vec{i} - 5\vec{j} + 6\vec{k}) \cdot [(-6\vec{i} + \vec{j} - 10\vec{k}) + (d_x\vec{i} + d_y\vec{j} + d_z\vec{k})] = 0$$

Performing the addition results

$$\vec{A} \cdot (\vec{B} + \vec{D}) = (4\vec{i} - 5\vec{j} + 6\vec{k}) \cdot [(-6 + k)\vec{i} + (1 + k)\vec{j} + (-10 + k)\vec{k}] = 0$$

$$\vec{A} \cdot (\vec{B} + \vec{D}) = [4(-6 + k) - 5(1 + k) + 6(-10 + k)] = 0$$

$$\vec{A} \cdot (\vec{B} + \vec{D}) = [(-24 + 4k) + (-5 - 5k) + (-60 + 6k)] = 0$$

$$\vec{A} \cdot (\vec{B} + \vec{D}) = (-89 + 5k) = 0$$

Therefore, the components of D are

$$k = 89/5$$

Chapter Four: Projectile Motion

1. Two projectiles are fired: one straight up with a speed of 100 m/s, and the other at an angle of 37^0 above the horizontal with an unknown speed. If the time of flights of the two projectiles are equal, (a) find the initial velocity of the second projectile, (b) find the maximum horizontal distance of the second projectile, and (c) find the maximum height each can rise.

Solution:

Fact #1: initial velocity of first projectile is v_{01}= 100 m/s,

Fact #2: initial velocity of second projectile is v_{02}= unknown,

Fact #3: angle of projection of second projectile is 37^0,

Fact #4: acceleration is $-10.0 m/s^2$,

Fact #5: time of flight is unknown,

Fact #6: each projectile will rise up until its vertical component of the velocity becomes zero, i.e. v_{1f}=0 is the velocity of first projectile at its max.

(a). While the first projectile describes one-dimensional motion; the second projectile is two-dimensional motion because it is fired at an angle. Thus, I can calculate the time it takes for the first projectile to reach its maximum height

$$v_{1f} = v_{01} + gt_{up}$$

Solving for t_{max} yields

$$t_{max} = -\frac{v_{01}}{g} = -\frac{100\,{}^{\mathrm{m}}/_{\mathrm{s}}}{\left(-10\,{}^{\mathrm{m}}/_{\mathrm{s}^2}\right)} = 10\mathrm{s}$$

Thus the time of flight (t_f) is twice the time it took to reach the maximum height, i.e.

$$t_f = 2t_{max} = 20\mathrm{s}$$

Since the second projectile is fired at an angle, one needs to resolve its initial velocity into horizontal and vertical components, i.e.

$$v_{02x} = \mathrm{v}_{02}\cos 37$$
$$v_{02y} = \mathrm{v}_{02}\sin 37$$

The second projectile will rise up until its initial vertical component of the velocity becomes zero, i.e.

$$v_{2f} = v_{0y} + at = 0$$

Where

$$v_{2f} = \mathrm{v}_0 \sin 37 - \mathrm{g}t_{\mathrm{max}} = 0$$

Solving for v_{02} yields,

$$v_{02} = \frac{gt_{max}}{\sin 37} = \frac{\left(10\,{}^{\mathrm{m}}/_{\mathrm{s}^2}\right)(10\mathrm{s})}{0.6} = \left({}^{1000}/_{6}\right){}^{m}/_{s} = 166.67\,{}^{m}/_{s}$$

(b) The horizontal range of the second projectile is the product of the horizontal component of the velocity and the time of flight, i.e.

$$x_{range} = v_{02}\cos 37\, \mathrm{t_f} = 166.67\,{}^{\mathrm{m}}/_{\mathrm{s}}\,(20\mathrm{s})\cos 37 = 2662.2\mathrm{m}$$

(c) The maximum height can be obtained by using one of the following general equations of motion.

$$\Delta x = \bar{v}t = \left(\frac{v + v_0}{2}\right)t$$
$$\Delta x = v_0 t + \left({}^{1}/_{2}\right)at^2$$
$$v^2 = v_0^2 + 2a\Delta x$$

Although any of the above equations must give the same result, one has to choose the equation that demands the least amount of effort. Selecting the first equation seems to make sense. For the first projectile, I have to make the following substitutions; $v=v_1=0$, $v_{01}=100$m/s, $t=t_{max}=10$s, $\Delta x=h_A=$unknown, then

$$h_A = \left(\frac{0 + 100\,{}^{\mathrm{m}}/_{\mathrm{s}}}{2}\right)(10\mathrm{s}) = 500\mathrm{m}$$

For the second projectile, I have to make the following substitutions $v=v_{f2y}=0$, $v_0=v_{02y}=$ [(1000)/6](m/s)sin37=100m/s, $t=t_{max}=10$s, $\Delta x=h_B=$unknown, then

$$h_B = \left(\frac{0 + 100\,{}^{m}/_{s}}{2}\right)(10s) = 500m$$

These calculations show that both projectiles rise to the same height.

2. **At what angle must a baseball be hit so that its maximum height is equal to the maximum horizontal distance it travelled on the baseball field?**

Solution:
Fact #1: Initial velocity of projectile is v_0=unknown,
Fact #2: angle of projectile is θ=unknown,
Fact #3: maximum height is y=unknown,
Fact #4: maximum horizontal distance is x=unknown,
Fact #5: time of flight is unknown.

The baseball describes two-dimensional motion, i.e. one horizontal and the other vertical. Even though there seems to be too many unknowns, I will apply the equations of motion and see if the problem will begin to unravel itself. The height is determined by the vertical component of the initial velocity and the acceleration due to gravity. The horizontal distance is determined by the horizontal component of the velocity. If I assign v_{ox} and v_{oy} to represent the horizontal and vertical component of the velocity then

$$v_{0x} = v_0 \cos\theta \qquad \text{and} \qquad v_{0y} = v_0 \sin\theta$$

The baseball will continue to rise up until the vertical component of the velocity becomes zero, i.e. $v_y=0$. Let t_{max} be the time for the baseball to reach its maximum height. Now I can find an expression for t_{max} by using one of the equations of motion, i.e.

$$v_y = v_{0y} + at$$

Upon substituting $v_y=0$, $v_{0y}=v_0\sin\theta$, $a=-g$, $t=t_{max}$ and rearranging I arrive at,

$$t_{max} = \frac{v_0 \sin\theta}{g}$$

The time of flight (t_f) of the baseball is twice the time to go up, i.e.

$$t_f = 2t_{max} = \frac{2v_0 \sin\theta}{g}$$

By using t_{max} and t_f, one can obtain the maximum height and maximum horizontal distance of the projectile, i.e.

$$y_{max} = v_0 \sin\theta\, t_{up} - \frac{1}{2} g t_{up}^2$$

$$x_{max} = v_0 \cos\theta\, t_f$$

Simplifying the above equations results in

$$y_{max} = \frac{v_0^2 \sin^2 \theta}{2g}$$

$$x_{max} = \frac{2v_0^2 \sin \theta \cos \theta}{g}$$

Equating the last two equation yields

$$\frac{v_0^2 \sin^2 \theta}{2g} = \frac{2v_0^2 \sin \theta \cos \theta}{g}$$

Eliminating v_0 and g and rearranging yields,

$$\tan \theta = 4$$

Thus the angle of projectile is obtained by taking the inverse tangent, i.e.

$$\theta = \tan^{-1} 4 = 75.96^0$$

3. The defense forces of a certain country are preparing to launch a missile against an enemy target, which is 100km from the launch site. If the gun is aimed at 37^0 above the horizontal, at what velocity must the missile be fired so that it hits the enemy target?

Solution:
Fact #1: horizontal distance is100km
Fact #2: angle of projectile is θ=37^0,
Fact #3: initial velocity is v_0= unknown.
This is similar to the baseball problem, which I just did. From the baseball problem, the following expression for the maximum horizontal distance was obtained, i.e.

$$x_{max} = \frac{2v_0^2 \sin \theta \cos \theta}{g}$$

Substituting x_{max}=100km, θ=37^0, and g=10m/s^2 yields,

$$v_0 = \sqrt{\frac{gx_{max}}{2 \sin \theta \cos \theta}} = \sqrt{\frac{\left(10\,{}^{m}/_{s^2}\right)(100{,}000\text{m})}{2 \sin 37 \cos 37}} = 1.02 \times 10^3 \text{m}$$

4. A man throws a ball horizontally with a speed of 20m/s across a pond, which is 100m wide. The ball was thrown horizontally from a height of 1.5m above the ground. The ball was subjected to a horizontally blowing wind, which resulted in a constant deceleration. If the ball reaches the other side of the pond with only a downward velocity, what was the deceleration of the ball?

Solution:
Fact #1: initial horizontal velocity is v_{0x}=20m/s,
Fact #2: angle of projectile is 0,
Fact #3: height of fall is y=1.5m,
Fact #4: acceleration in the y-direction is a_y=g,

Fact #5: acceleration in the x-direction is a_x=unknown

Fact #6: maximum horizontal distance is 100m,

Fact #7: initial vertical component velocity is $v_{0y}=0$,

Fact #8: final horizontal velocity is $v_x=0$,

Fact# 9: time of flight is t=unknown.

By applying the following equation of motion for the vertical part, I can obtain the time of flight of the ball, i.e.

$$y = v_{0y}t + (1/2)a_y t^2$$

Solving for t yields,

$$t = \sqrt{\frac{2y}{g}} = \sqrt{\frac{2(1.5m)}{10\,m/s^2}} = 0.55s$$

Now I can find the acceleration in the horizontal direction by using;

$$a_x = \frac{\Delta v}{\Delta t} = \frac{0 - 20\,m/s}{0.55s} = -36.51\,m/s^2$$

5. A battleship simultaneously fires two shells with the same muzzle velocity at an enemy command and communication complex that lies on the same line of sight at known distances from the battleship. If the centers are 120.557 km apart: (a) which complex will get hit first, (b) what tactical advantage does this outcome have for the forces commanding the battleship, and (c) in order to hit the farther target, the angle must be 45 degrees and to hit the nearer target the angle must be 60 degrees, what fraction of time longer does it take to hit the nearer target?

Solution:

Fact #1: initial velocity=muzzle velocity= v_0=unknown,

Fact #2: distance of closer target is x=unknown,

Fact #3: distance of farther target is x+120,577m

Fact #4: angle of missile targeting closer site is θ_1=unknown

Fact #5: angle of missile targeting farther site is θ_2=unknown

My intuition tells me that the closer target will get hit first, but I have learned not to relay on intuition. I would rather workout the problem and let my calculations speak for me. I think Professor John will be happy to know about my approach.

In the baseball problem (Problem #3) I found that the time of flight of a projectile was given by,

$$t_f = \frac{2v_0 \sin\theta}{g}$$

Let t_{f1} and t_{f2} be the time-of-flights it takes the projectile to reach their targets respectively,

$$t_{f1} = \frac{2v_0 \sin\theta_1}{g}$$

$$t_{f2} = \frac{2v_0 \sin\theta_2}{g}$$

Taking the ratio of t_{f1} to t_{f2} yields,

$$\frac{t_{f1}}{t_{f2}} = \frac{\sin\theta_1}{\sin\theta_2}$$

The velocity drops out since the muzzle velocities are the same. In order to hit the targets, the missiles must be fired at angles such that θ_1 is greater than θ_2. As a result, the ratio is

$$\frac{t_{f1}}{t_{f2}} = \frac{\sin\theta_1}{\sin\theta_2} > 1$$

This means t_{f2} is smaller than t_{f1}, hence the farther target will get hit first before the closer target, contrary to my intuition.

iii. The tactical advantage of destroying the farther complex first, will be the disruption of communications used to warn the leaders of the enemy of the impending attack and consequently preventing them from retreating to their bunkers.

iv. Since the horizontal component of the velocity of each projectile is the same, the ratio of t_{f1} to t_{f2} yields

$$\frac{t_{f1}}{t_{f2}} = \frac{\sin 60}{\sin 45} = \sqrt{3/2} = 1.22$$

This means it takes 1.22 times longer to hit the closer target.

6. A radar system has observed an unidentified object falling from an altitude of 5,500m towards a depot housing important property that is located 20 km away from the radar station. The radar showed the object's speed was 18 km/h. In an effort to destroy the falling object in air, the observers decide to fire a missile immediately with a muzzle velocity of 1000m/s. (a) At what angle should the missile be fired to hit the falling object at least 500m before it reaches the depot, and (b) does the missile hit the object as it is rising or falling?

Solution:
Fact #1: horizontal distance is 20km=20,000m,
Fact #2: initial velocity of missile is 1,000m/s,
Fact #3: acceleration is g
Fact #4: allowed distance of fall is 5km=5,000m
Fact #5: initial angle of missile is unknown,
Fact #6: time of fall is unknown,
Fact# 7: object's initial speed is =18km/h=5m/s.

This looks like a tall order. I am not sure whether I can do this without Professor John's help, but I am going to use every ounce of my talent to tackle it. Let me find out how long it takes the object to fall within 500m of the depot. This means the object has fallen 5,000m. I can use the following equation of motion to find the time, i.e.

$$y = v_0 t + \frac{1}{2} g t^2$$

Taking the downward direction as positive and substituting the known facts yield,

$$5{,}000\text{m} = (5\,{}^{\text{m}}/_{\text{s}})\text{t} + \frac{1}{2}\left(10\,{}^{\text{m}}/_{\text{s}^2}\right)\text{t}^2$$

Simplifying yields

$$5\text{t}^2 + 5\text{t} - 5{,}000 = 0$$

Solving this by using the quadratic formula yields,

$$\text{t} = \frac{-1 \pm \sqrt{1^2 - 4(1)(-1{,}000)}}{2(1)} = \frac{-1 \pm 63.25}{2}$$

$$\text{t} = 31.12\text{s} \quad \text{or} \quad \text{t} = -32.125\text{s}$$

Since there is no negative time, only t=31.12s is acceptable. This means the missile must meet the object in 31.12s. I know that the horizontal component of the initial muzzle velocity is responsible for taking the missile towards the falling object, i.e.

$$\text{x} = \text{v}_0 \cos\theta\, \text{t}$$

Substituting x, v_o and t and solving for θ yields,

$$\theta = \cos^{-1}\left(\frac{20{,}000m}{\left(1000\frac{m}{s}\right)(31.12s)}\right) = 50^0$$

Thus, the missile must be fired at an angle of 50 degrees above the horizontal so that it can hit the falling unknown object. I can decide whether the missile hits the object on its way up or down by finding the time it takes to rise to the maximum height. As we saw earlier, the time to reach the maximum height is calculated by using,

$$\text{t}_{\text{max}} = \frac{\text{v}_0 \sin\theta}{\text{g}} = \frac{(1000\,{}^{\text{m}}/_{\text{s}})\sin 50}{10\,{}^{\text{m}}/_{\text{s}^2}} = 76.6\text{s}$$

Since this is greater than the time the missile takes to reach the object, it means the missile hits the object on its way up.

7. The operators of a cannon are preparing to fire cannonballs with unknown muzzle velocity from a gun aiming 60 degrees above the horizontal. The operators of the gun noticed that there is a 100m tall building located 450 m in front of them. (a) At what muzzle velocity must the gun be fired so that the cannonballs will clear the building, and (b) where will the cannonballs hit the ground?

Solution:
Fact #1: horizontal distance of building is 450m,
Fact #2: height that must be cleared is 100m,
Fact #3: acceleration is g (downwards)
Fact #4: angle is 60 degrees
Fact #5: muzzle velocity of projectile is v_0=unknown,

Fact #6: horizontal component of the velocity is $v_0\cos\theta$=unknown,

Fact #7: initial vertical component of the velocity is $v0\sin\theta$=unknown,

This seems like another tall task again, but I am up to it this time. Let me begin by considering the horizontal distance, i.e.

$$x = v_0 \cos\theta\, t$$

Solving for time yields,

$$t = \frac{x}{v_0 \cos\theta} = \frac{450\,{}^{m}/_{s}}{v_0 \cos 60} = \frac{900}{v_0} \qquad (1)$$

The vertical component of the position is given by

$$y = v_0 \sin\theta\, t - \frac{1}{2}gt^2$$

Substituting the value of time obtained in Equation (1) gives;

$$y = (v_0 \sin 60)\left(\frac{900}{v_0}\right) - \frac{1}{2}g\left(\frac{900}{v_0}\right)^2 \qquad (2)$$

The minimum value of y that must be cleared is 100m. Putting this value into the above equation and solving for v_0 yields,

$$100 = 779.42 - \frac{4050000}{v_0^2}$$

Hence

$$v_0 = \sqrt{(4050000)(0.00147184363)} = 77.21\,{}^{m}/_{s}$$

(b) The range of the projectile is obtained by using

$$R = \frac{v_0^2 \sin 2\theta}{g} = \frac{(77.21\,{}^{m}/_{s})^2 \sin 120}{10\,{}^{m}/_{s^2}} = 516.24\text{m}$$

8. A college student was firing practice-rounds from the edge of the roof of an 80m tall building. Her gun can fire a bullet with a muzzle velocity of 100 m/s. The student set up the gun horizontally. As she was firing the bullet shells were falling off the edge of the roof. (a) Between the bullet and the shell, which one would reach ground first, (b) how long does it take each to hit the ground, (c) how far away from the building will the bullet hit the ground, and (d) with what speed will the bullet hit the ground?

Solution:

Fact #1: height to fall is 80m,

Fact #2: initial vertical component of velocity of shells is 0,

Fact #3: acceleration is g (downwards)

Fact #4: initial vertical component of velocity of bullet is 0

Fact #5: initial horizontal component of velocity of shell is 100m/s,

Fact #6: angle of bullet is 0,

(a). I think this problem is similar to the packages dropped from an airplane and a stationary helicopter that we discussed in the chapter on projectile motion. Since the time of descent depends on the initial vertical component of the velocity, both the bullet and shell will hit the ground at the same time.

(b). The time is obtained by using

$$y = v_0 \sin\theta\, t - \frac{1}{2}gt^2$$

Solving for t gives

$$t = \sqrt{\frac{2y}{g}} = \sqrt{\frac{2(80m)}{10\, ^m/_{s^2}}} = 4s$$

(c). The horizontal distance is calculated from,

$$x = v_x t = 100\, ^m/_s\, (4s) = 400m$$

(d). The horizontal component of the velocity is constant, i.e.

$$v_x = 100\, ^m/_s$$

The vertical component is obtained from

$$v_y = v_0 \sin 0 - gt = 0 - \left(10\, ^m/_{s^2}\right)(4s) = -40\, ^m/_s$$

Thus the speed with which the bullet hits the ground is;

$$v = \sqrt{v_x^2 + v_y^2} = \sqrt{(100\, ^m/_s)^2 + (-40\, ^m/_s)^2} = 107.7\, ^m/_s$$

9. A package was dropped from an airplane while it was rising. The package was dropped from an altitude of 200m. If the plane's front end was pointing at 37^0 above the horizontal and its speed was 100m/s, (a) describe the motion of the package, (b) how long does the package take to reach the ground, and (c) where will the package land?

Solution:

Fact #1: height from which the package was dropped is 200m,

Fact #2: initial velocity of package is100m/s,

Fact # 3: Angle of initial velocity is 37 degrees,

Fact # 4: acceleration is g (downwards),

Fact #5: initial vertical component is v_{0y}=(100m/s)sin37=60m/s,

Fact #6: initial horizontal component is v_{0x}=(100m/s)cos37=80m/s,

Since the package was dropped while the plane was ascending, it will continue to rise up until its vertical component of the velocity becomes zero. Considering physical quantities pointing upwards as positive and downwards as negative, I can apply the following equation of motion to solve for the time it takes for the package to hit the ground, i.e.

$$y = v_{0y}t + \frac{1}{2}at^2$$

Substituting the values given yields,

$$-200\text{m} = 60\text{t} + \frac{1}{2}\left(-10\,{}^{m}/_{s^2}\right)\text{t}^2$$

Simplifying further gives

$$\text{t}^2 - 12\text{t} - 40 = 0$$

Solving for t yields

$$\text{t} = \frac{12 \pm \sqrt{(12)^2 - 4(1)(-40)}}{2}$$

Hence

$$\text{t} = 14.72\text{s} \qquad \text{or} \qquad \text{t} = -2.72\text{s}$$

The negative time is discarded because there is no negative time; hence the package will take 14.72 seconds to hit the ground.

c. The horizontal distance is obtained by using,

$$\text{x} = \text{v}_{\text{ox}}\text{t} = (80\,{}^{\text{m}}/_{\text{s}})(14.72\text{s}) = 1177.6\text{m}$$

10. A football punter kicks a ball from ground level at an angle of 37^0 above the horizontal towards the opposing teams' field. If the ball was observed to have a speed of 25m/s when it was at its highest point, (a) what was the speed with which it was kicked, (b) what will be the time of flight of the football and (c) how far does the ball travel downfield?

Solution:
Fact #1: angle of projectile is θ=37 degrees,
Fact #2: initial velocity is v_0=unknown,
Fact #3: velocity at the highest point is 25m/s,
Fact #4: acceleration is g,
Fact #5: horizontal distance is x=unknown.

Finally, I got an easier problem to deal with this time. At the highest point, the velocity is completely horizontal. Since the horizontal velocity is constant throughout the football's motion, it is related to the initial velocity by the following equation

$$\text{v}_{\text{ox}} = \text{v}_0 \cos\theta = 20\,{}^{\text{m}}/_{\text{s}}$$

Solving for v_0 yields,

$$\text{v}_0 = \frac{20\,{}^{\text{m}}/_{\text{s}}}{\cos 37} = 31.25\,{}^{\text{m}}/_{\text{s}}$$

The time of flight is obtained by using,

$$\text{t}_\text{f} = \frac{2\text{v}_0 \sin\theta}{\text{g}} = \frac{2(25\,{}^{\text{m}}/_{\text{s}}) \sin 37}{\left(10\,{}^{\text{m}}/_{\text{s}^2}\right)} = 3.0\text{s}$$

The horizontal distance is obtained by

$$x = (v_0 \cos\theta)t_f = (25.0\, {}^{m}/_{s}) \cos 37\, (3.0s) = 60.2m$$

Additional Tidbits

1) If a projectile is fired at an angle of 37^0 above the horizontal with initial speed of 100 m/s on a flat ground, what is the magnitude and direction of the projectile's acceleration when the projectile is (a) before the highest point, (b) at its highest point, and (c) after its highest point?

2) If a projectile is fired at an angle of 37^0 above the horizontal with initial speed of 100 m/s on a flat ground, what is its velocity when it reaches its highest point?

Answers' to Tidbits

1) The acceleration due to gravity is the same no matter where the object is, i.e. $9.8m/s^2$. Hence the answer to all questions is the same.
2) Even though the vertical component of the velocity is zero, its horizontal velocity is nonzero. Hence the velocity at the highest point is

$$v_x = v_0 \cos\theta = (100\, {}^{m}/_{s}) \cos 37 = 80\, {}^{m}/_{s}$$

Chapter Five: Force and Motion

1. Drawing a free body diagram means…

(a) Balancing the forces acting on each object.

(b) Identifying the forces acting on all objects simultaneously.

(c) Identifying the forces on each object one object at a time.

(d) Finding the net force on each object.

Solution:

Drawing the free body diagram on an object is the same as identifying all the forces acting on the object. Hence choice (c) is the correct answer.

2. Which of the following physical quantities have the same direction all the time?

(a) Displacement and velocity. **(b) Velocity and acceleration.**

(c) Acceleration and net force. **(d) Answers (a) and (c).**

Solution:

I know an object can only be displaced in the direction of its motion. Thus velocity and displacement have the same directions at all times. The acceleration of an object has the same direction as its velocity and displacement only when the object is moving in one direction and its velocity is increasing. The acceleration of an object has opposite direction to its velocity and displacement only when the object is moving in one direction and its velocity is decreasing. Velocity and displacement are perpendicular to the acceleration when the object is moving on a circular path. The acceleration of an object and the net force have the same direction at all times. Thus choice (d) is the right answer.

3. When you are a passenger in a bus, and the bus suddenly brakes, you tend to lean forward. Describe what concept explains this reaction.

Solution:
According to Newton's first law every object has inertia, which means that an object resists change in its current state of motion. When the bus brakes, a force is applied opposite to the direction of its motion. According to Newton's first law, a person inside the bus tends to continue to move at the same velocity as that before the bus applied its brakes. The person resisting the change applies a force, which produces a forward leaning motion.

4. How is the direction of frictional force related to the direction of motion?

(a) It is always opposite to the direction of motion.
(b) It is the same direction as motion.
(c) It has the same direction as the external force at all times.
(d) It is opposite in direction to the net external force at all times.

Solution:
The direction of frictional forces is always opposite to the direction of motion. Thus, choice (a) is the correct answer.

5. Which of the following is true about the normal force?

(a) It is a reaction force exerted on an object that is not in contact with another firm object.
(b) It is a reaction force exerted on an object that is in contact with another firm object.
(c) It is a force that is always equal to the weight of an object.
(d) It is a force that is always greater than the weight of an object.
(e) One's objects normal force is the opposite of the other objects normal force.

Solution:
The normal reaction force is the result of two surfaces pressing against each other. In order for the two surfaces to press against each other, they must be in contact with each other. Thus, choice (b) and (e) are the right answer.

6. How do you describe the motion of an object when the vector sum of all the forces acting on the object is zero?

(a) The object cannot be moving at all.

(b) The object is moving with uniform velocity.

(c) The object is moving with constant acceleration.

(d) The object's displacement is zero.

Solution:

According to Newton's second law when the net force on an object is zero, it means the object is not accelerating. However, the object can still be moving at constant (uniform) velocity. Hence choice (b) is the correct answer.

7. A 10 kg block of mass is sliding on a horizontal surface with an initial speed of 6.0 m/s. The block travels a distance of 5 m as it slows to a stop. Calculate the magnitude of the braking force responsible for stopping the block.

Solution:

Fact #1: mass is 10kg,

Fact #2: initial velocity is v_0=6.0m/s,

Fact #3: final velocity is v=0,

Fact #4: displacement is Δx=5.0m,

Fact #5: Acceleration is a =unknown.

The net force on an object is related to its acceleration by Newton's second law of motion, i.e.

$$\vec{F} = m\vec{a}$$

In order to obtain the net force, it is obvious that I have to find the acceleration by using the equations of motion from chapter 2. An equation to use is:

$$v^2 = v_0^2 + 2a\Delta x$$

Solving for the acceleration yields

$$a = \frac{v^2 - v_0^2}{2\Delta x} = \frac{0 - (6\,{}^{m}/_{s})^2}{2(5m)} = -3.6\,{}^{m}/_{s^2}$$

Thus, the magnitude of the braking force is

$$|\vec{F}| = m|\vec{a}| = (10kg)\left(3.6\,{}^{m}/_{s^2}\right) = 36N$$

8. A 100 kg block is resting on a frictionless surface. If the two forces act on the block as indicated, and the angle is $\theta = 37^0$.

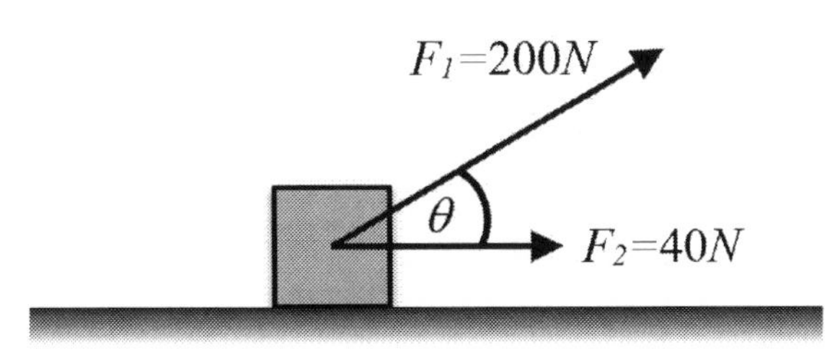

(a) Draw the block's free body diagram, (b) find the acceleration of the block, and (c) find the normal reaction force exerted by the ground on the block.

Solution:
Fact #1: $F_1 = 200N$,
Fact #2: orientation of F_1 is $\theta = 37^0$,
Fact #3: $F_2 = 40N$,
Fact #4: The normal reaction force is F_N (upwards).
Fact #5: the weight of the block is W=mg (downwards).

(a) This is a one-dimensional problem. Before I proceed to solve the problem, I must first identify all the forces acting on the block. In other words, I must draw the block's free body diagram. Now I can draw the free body diagram as shown below.

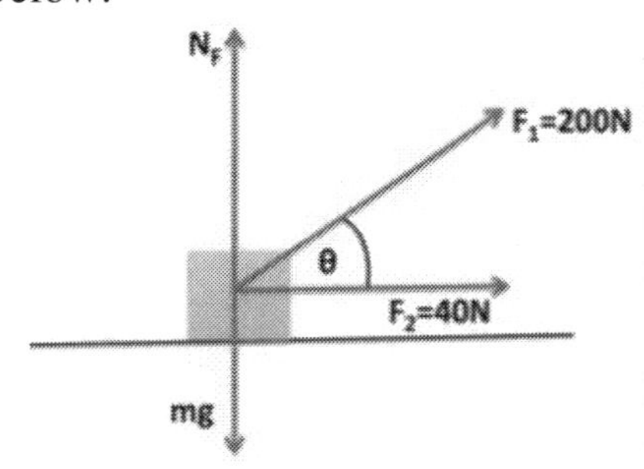

(b) Here I have to find the net horizontal force because when it comes to accelerating the block it is the only force that causes it. To do so I have to resolve each force into horizontal and vertical components.

Force	horizontal component	vertical component
F_1	200Ncos37=160N	200Nsin 37=120N
F_2	40N	0
F_N	0	F_N
W	0	-mg

The next step is to add the components separately. If I represent the net horizontal and vertical component of forces by F_x and F_y respectively, then

$$F_x = 160N + 40N = 200N$$

$$F_y = 120N + F_N - mg$$

By applying Newton's second law to each component, I can write

$$F_x = ma_x = 200N$$

The horizontal acceleration is obtained by dividing the net force by the mass, i.e.

$$a_x = \frac{200N}{100kg} = 2\,{}^{m}/_{s^2}$$

The vertical component of force is

$$F_y = ma_y = 120N + F_N - mg$$

where a_x and a_y are the horizontal and vertical accelerations respectively. Since the motion is restricted to the horizontal direction, then

$$a_y = 0$$

(c) Substituting this value of a_y into the above equation yields

$$F_N = mg - 120N = (100kg)\left(10\,{}^{m}/_{s^2}\right) - 120N = 880N$$

9. An object of 10kg is placed on a frictionless inclined plane as shown below. If θ_1=45^0 and θ_2=15^0, find (a) the necessary force to make the object move up the incline at constant velocity, and (b) the normal force exerted by the inclined plane.

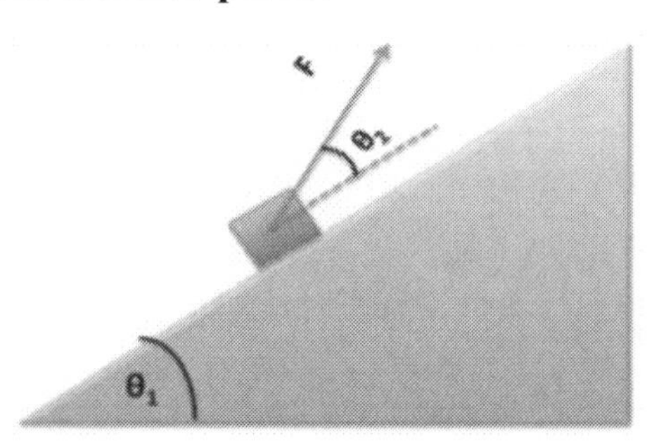

Solution:

Fact # 1: The applied force is F

Fact #2: The weight of the block straight down is W=mg,

Fact #3: The normal force exerted by the inclined plane is F_N,

Fact #4: The angle of inclined plane is θ_1=45^0,

Fact #5: The angle of applied force is θ_2=15^0.

Once again I have to start by identifying all the forces acting on the block. Thus the free body diagram is as shown below.

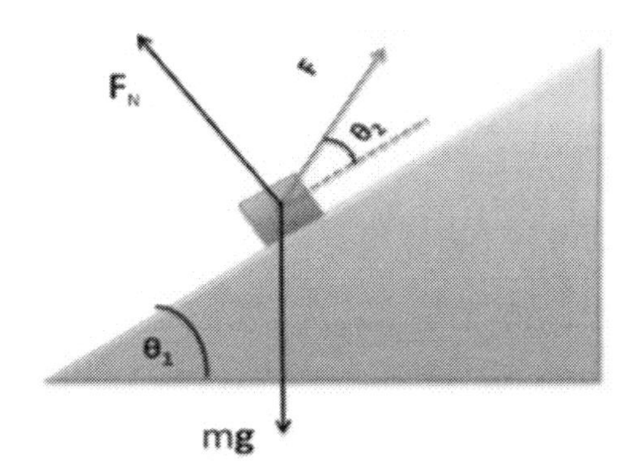

The next step is to resolve each force into parallel and perpendicular components as indicated below.

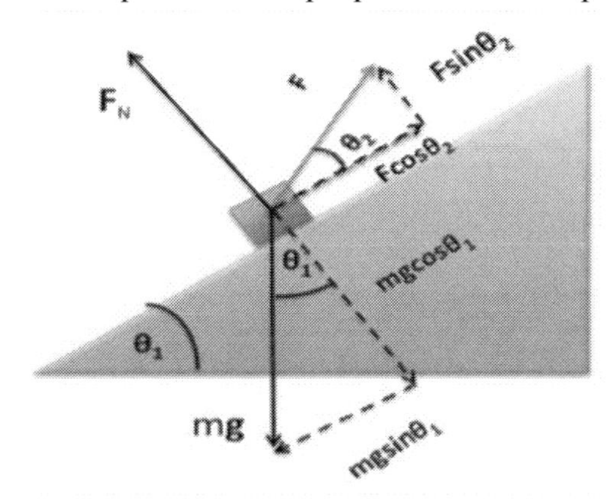

Force	parallel component	Perpendicular component
F	$F\cos\theta_2$	$F\sin\theta_2$
W	$-mg\sin\theta_1$	$-mg\cos\theta_1$
F_N	0	F_N

(a) Now I can add the forces component by component. Let $F_{\parallel}$ and $F_{\perp}$ represent the sum of the parallel and perpendicular components respectively, i.e.

$$F_{\parallel} = F\cos\theta_2 - mg\sin\theta_1 = ma_{\parallel}$$

$$F_{\perp} = F\sin\theta_2 - mg\cos\theta_1 + F_N = ma_{\perp}$$

where $a_{\parallel}$ and $a_{\perp}$ are the parallel and perpendicular components of the acceleration. Since the motion of the block is restricted to be along the inclined plane, there is no perpendicular acceleration, i.e.

$$a_{\perp} = 0$$

Since the object is moving at constant velocity, the parallel acceleration is also zero, i.e.

$$a_{\parallel} = 0$$

Substituting these values of the acceleration into the above equation yield,

$$F\cos\theta_2 = mg\sin\theta_1 \qquad (1)$$

$$F_N = mg - F\sin\theta_2 \qquad (2)$$

Solving for F from Eq. (1) gives

$$F = \frac{mg \sin \theta_1}{\cos \theta_2} = \frac{(100kg)\left(10\,{}^{m}/_{s^2}\right) \sin 45}{\cos 15} = 732.1N$$

(b) Now I can calculate the normal force by using Eq. (2) and the newly obtained value of F, i.e.

$$F_N = mg - F \sin \theta_2 = (100kg)\left(10\,{}^{m}/_{s^2}\right) - (732.1N \sin 15) = 810.5N$$

10. An applied force is lifting two objects that are connected by an inextensible rope as shown in the diagram below.

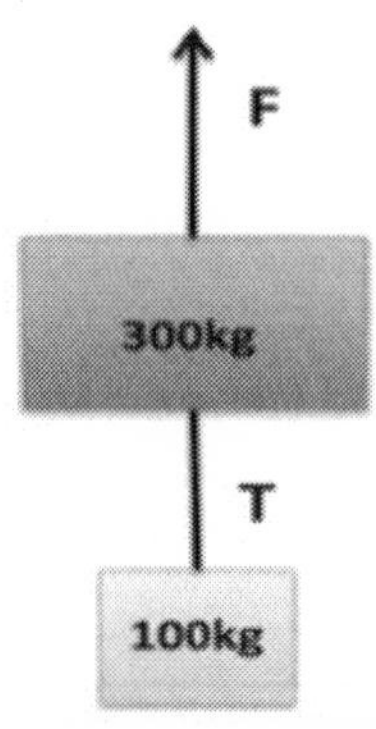

If the acceleration of the blocks is 5.0 m/s^2, (a) find the applied force and (b) find the tension T in the rope between the two blocks.

Solution:
As usual, in these types of problems one has to begin by identifying the forces acting on each object. The forces acting on the 100 kg block are the tension (T) in the rope and its weight (W_1). The forces on the 300 kg block are the external force (F), its weight (W_2) and the tension (T). Thus the free body diagram of each block can be represented as shown below.

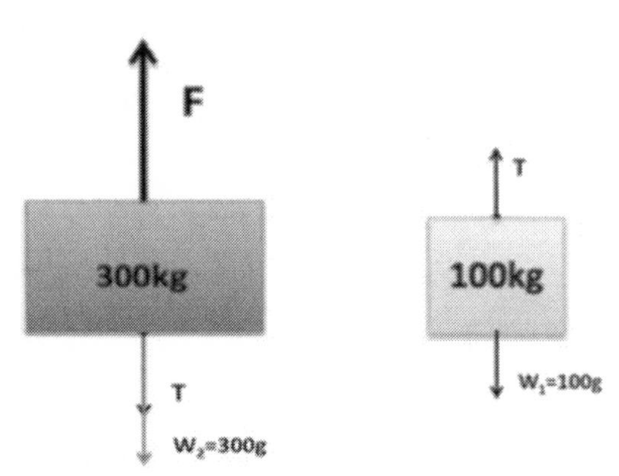

The next step is to find the net force on each block. Let F_1 and F_2 be the net force on each block respectively where,

$$F_1 = T - 100g = 100a \qquad (1)$$

$$F_2 = F - T - 300g = 300a \qquad (2)$$

Since the acceleration is given, I can find the value of the tension T from the first equation, i.e.

$$T = 100a + 100g = (100kg)\left(5\,{}^{m}/_{s^2} + 10\,{}^{m}/_{s^2}\right) = 1500N$$

The applied force (F) is obtained by substituting the newly found value of the tension into the second equation, i.e.

$$F = 300a + T + 300g = 1500N + 1500N + 3000N = 6000N$$

11. A hockey puck is sitting idle when 5 Lincoln Stars players approach. The Players all arrive and hit the puck simultaneously. Player 1 hits with 5 N, player 2 hits with 3 N at an angle of 30° with respect to player 1, player 3 hits with 4 N at an angle of -120° with respect to player 2, player 4 hits with a force of 5 N at an angle of 150° with respect to player 1, and player 5 hits with 2 N at -60° with respect to player 4.

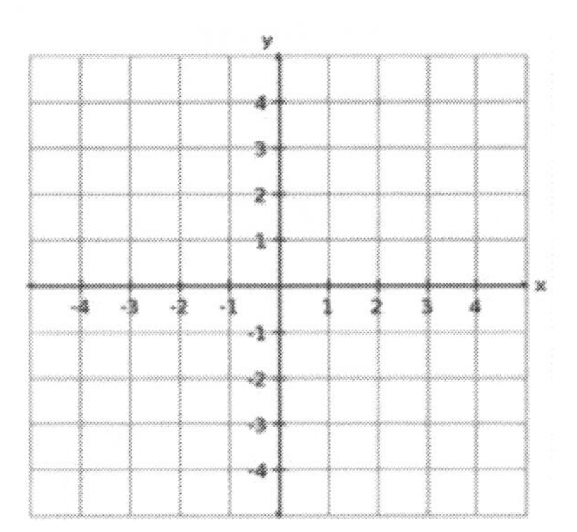

(a) Draw the force vectors acting on the puck due to each player's hit, and (b) what is the net force on the puck?

Solution:

Suppose the forces applied by the different players are represented by F_1, F_2, F_3, F_4 and F_5 respectively. For convenience let F_1 lie along the x-axis. Then F_2 will be in the first quadrant, F_3 will be on the negative y-axis, F_4 will be in the third quadrant at 150^0 from the line of F_1 and F_5 will lie along the y-axis as shown in the free body diagram below.

Note: Angles measured counterclockwise are taken as positive and angles measured in the clockwise direction are taken as negative.

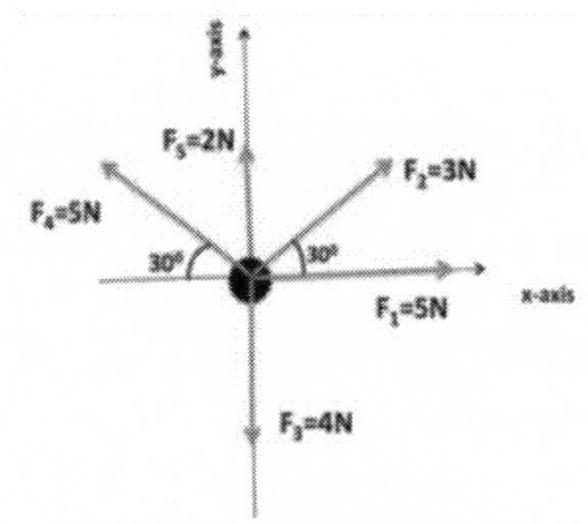

b) The net force is obtained by the analytical approach used before, i.e.

Force	x-component	y-component
F_1	5N	0
F_2	3Ncos30=2.6N	3Nsin30=1.5N
F_3	0	-4N
F_4	-5Ncos30=-4.33N	5Nsin30=2.5N
F_5	0	2N

Now I have to add the x and y components separately. If F_x and F_y are the sum of the x and y components respectively then,

$$F_x = 5N + 2.6N - 4.33N = 3.27N$$

$$F_y = 1.5N - 4N + 2.5N + 2N = 2.0N$$

The magnitude of the net force is obtained using Pythagoras' theorem,

$$F = \sqrt{(3.27N)^2 + (2.0N)^2} = 3.83N$$

The direction of the net force is found by taking the inverse tangent of the ratio of the y-component to the x-component, i.e.

$$\theta = \tan^{-1}\left(\frac{F_y}{F_x}\right) = \tan^{-1}\left(\frac{2}{3.27}\right) = 31.45 \text{ deg}$$

12. A force of 100 N presses a 3 kg block as indicated below. The coefficient of static friction between the block and the wall is μ_s = .50. Find the force on the block exerted by the wall.

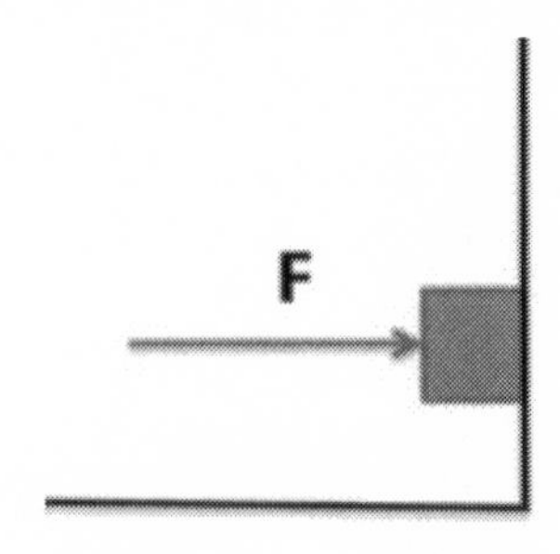

Solution:

As usual I should identify the forces acting on the block. The forces acting on the block are the applied force (**F**) as shown, the weight of the block (mg) downward, the frictional force (**f**) upwards and the force exerted by the wall (**N_F**). Thus I have the following facts:

Fact #1: applied force (F)= 100N to the right,

Fact #2: coefficient of friction (μ_s)=0.5,

Fact #3: frictional force (**f**)=unknown,

Fact #4: Normal force (**N_F**)= unknown.

The free body diagram of the block will be as follows.

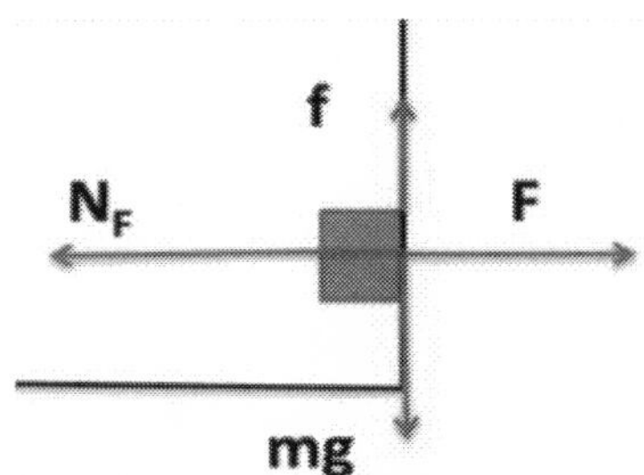

Force	horizontal component	vertical component
F	100N	0
f	0	f
mg	0	-mg
N_F	$-N_F$	0

The forces in the horizontal direction are balanced which means,

$$100N - N_F = 0 \Rightarrow N_F = 100N$$

The frictional force is given by

$$f = \mu_s N_F = (0.5)100N = 50N$$

The force exerted by the wall is

$$\vec{F}_w = -(100N)\vec{\imath} + (50N)\vec{\jmath}$$

Thus the magnitude of the force exerted by the wall is given by

$$F_w = \sqrt{(N_F)^2 + f^2} = \sqrt{(100N)^2 + (50N)^2} = 111.8N$$

Chapter Six: Work and Kinetic Energy

1. What is the kinetic energy of a stationary 10kg object?

Solution:
Fact #1: mass is10 kg,
Fact #2: velocity is 0,
The kinetic energy is the energy of its motion. Since the object is motionless, its kinetic energy is zero.

2. What does the change in the kinetic energy of an object equal to?
  (a) The work done by each force that is acting on the object.
  (b) The work done by the net force acting on the object.
  (c) The total work done by all the forces that have the same direction.
  (d) None of the above.

Solution:
According to the work-kinetic energy theorem, the change in the kinetic energy of an object represents the total work done on the object. Thus choice (b) is the correct answer.

3. A block of mass 20 kg that is sliding on a horizontal surface with an initial speed of 6.0 m/s travels a distance of 5 m before it slows to a stop. (a) Calculate the work done by the braking force responsible to stop the block, and (b) find the magnitude of the stopping force.

Solution:
This is easy. I know that net work on an object is equivalent to the change in the kinetic energy of the object.
Fact #1: mass of block is 20 kg,
Fact #2: initial velocity is v_0=6.0m/s,
Fact #3: final velocity is 0,
Fact #4: distance is 5m,
By definition the work done by a constant force acting on an object is given by

$$w = Fd\cos\theta \qquad (1)$$

where F is the magnitude of the force, d is magnitude of the displacement and θ is the angle between the directions of the force and displacement. In the present problem, the force is opposing the motion of the object; hence,

$$\theta = 180 \text{ degrees}$$

Since the force is unknown, I cannot calculate the work done by using the above formula. Thus, the only way to find the work done is to use the equivalence of work and change in kinetic energy, i.e.

$$W = \Delta K = K_f - K_i$$

Where K_f and K_i are the final and initial kinetic energies respectively, i.e.

$$K_i = \frac{1}{2}mv_0^2 = (0.5)(10kg)(6\,{}^{m}/_{s})^2 = 360J$$

$$K_f = \frac{1}{2}mv_f^2 = (0.5)(10kg)(0)^2 = 0$$

Thus the change in the kinetic energy of the object is:

$$\Delta K = K_f - K_i = 0 - 360J = -360J$$

Hence the work done by the opposing force is:

$$W = \Delta K = -360J$$

To find the force, I will use the newly found work in Equation 1

$$Fd\cos 180 = W = -360J$$

Solving for F yields

$$F = \frac{-360J}{-5m} = 72N$$

4. In the figure below, a block of mass M is moving down an inclined plane. Its initial velocity is v_0, and the height of the ramp is h. There is a kinetic frictional force f_k between the block and the ramp.

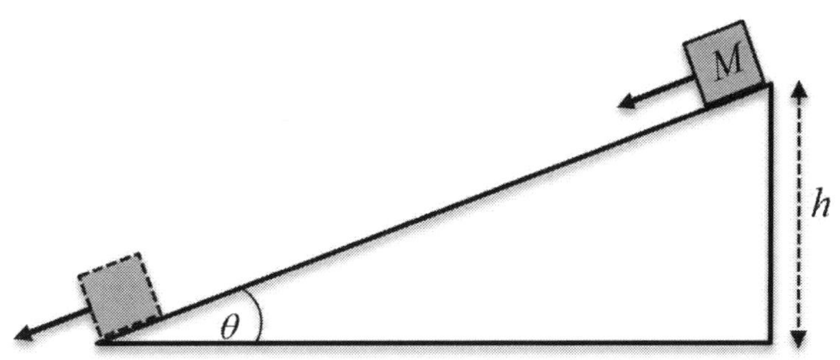

(a) Find the work done by the force of gravity, (b) find the work done by friction, and (c) find the net work done on the object.

Solution:

a) The block is moving down the plane due to the component of its weight in that direction. The component of the weight parallel to the plane is

$$F_{\parallel} = Mg\sin\theta$$

If I let the displacement down the plane as positive, then the work done by gravity is

$$W_g = (\mathrm{Mg}\sin\theta)L$$

Substituting

$$\mathrm{h} = \mathrm{L}\sin\theta$$

gives

$$\mathrm{W_g} = \mathrm{Mgh}$$

b) The work done by the frictional force is given by

$$\mathrm{W_f} = -\mathrm{f_k L}$$

The work done by friction is negative because the frictional force acts opposite to the direction of motion, which decreases its kinetic energy.

The net work done on the object is the same as the change in its kinetic energy. The initial kinetic energy is given by

$$\mathrm{K_i} = \frac{1}{2}\mathrm{Mv_0^2}$$

Next I have to find the final velocity of the object at the bottom of the ramp. One approach is to use find the acceleration down the incline by applying Newton's 2nd law.

$$F_{net,\parallel} = \mathrm{Mg}\sin\theta - \mathrm{f_k} = \mathrm{Ma_\parallel}$$

Solving for the acceleration down the plane yields

$$a_\parallel = \frac{\mathrm{Mg}\sin\theta - \mathrm{f_k}}{M}$$

Now I have to use the equation of motion to find the final speed down the ramp.

$$\mathrm{v^2} = \mathrm{v_0^2} + 2\mathrm{a_\parallel L}$$

Hence

$$\mathrm{v^2} = \mathrm{v_0^2} + 2\left(\frac{\mathrm{Mg}\sin\theta - \mathrm{f_k}}{\mathrm{M}}\right)\mathrm{L}$$

The corresponding final kinetic energy is

$$\mathrm{K_f} = \frac{1}{2}\mathrm{Mv^2} = \frac{1}{2}\mathrm{M}\left(\mathrm{v_0^2} + 2\left(\frac{\mathrm{Mg}\sin\theta - \mathrm{f_k}}{\mathrm{M}}\right)\mathrm{L}\right)$$

Thus the net work done on the block is

$$\mathrm{W} = \mathrm{K_f} - \mathrm{K_i} = 2\left(\frac{\mathrm{Mg}\sin\theta - \mathrm{f_k}}{\mathrm{M}}\right)\mathrm{L} \qquad (1)$$

5. For Problem #4, compute the net work done on the block, for M=5kg, L=10m, the coefficient of kinetic friction μ_k=0.2 and θ=30^0.

Solution:

Before I can apply Equation (1) to compute the net work done on the block, I must find the magnitude of the frictional force. The magnitude of the frictional forces is obtained by using

$$\mathrm{f_k} = \mu_\mathrm{k}\mathrm{N} \qquad (2)$$

The normal force is calculated as follows;

$$\mathrm{N} = \mathrm{mg}\cos\theta = (5\mathrm{kg})\left(10\,{}^{\mathrm{m}}/_{\mathrm{s^2}}\right)\cos 30 = 43.3\mathrm{N} \qquad (3)$$

The frictional force is then

$$f_k = \mu_k N = (0.2)(43.35\text{N}) = 8.66\text{N}$$

Thus the net work done is calculated as

$$W = 2\left(\frac{Mg\sin\theta - f_k}{M}\right)L = 2\left(\frac{(5kg)\left(\left(10\,{}^{m}/_{s^2}\right)\sin 30 - 8.66N\right.}{5kg}\right)(10m) = 65.36\text{J}$$

6. A block of mass 5kg slides up an inclined ramp with an angle of θ=53⁰ above the horizontal. The block begins moving up the ramp with an initial velocity of 20m/s. If the block comes to a stop at 5m up the ramp as shown in the figure;

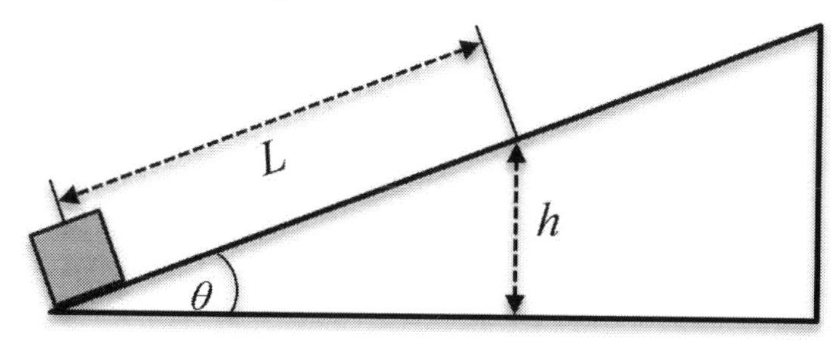

(a) Is kinetic friction present between the block and the ramp? Justify your answer, and (b) how much work is done by the kinetic friction on the block between its starting point and the point where it came to rest?

Solution:
I have to calculate the initial kinetic energy of the block and the final gravitational potential energy. If the final potential energy is less than the initial kinetic energy then it means friction is present.
Fact #1: initial speed is 20m/s,
Fact #2: mass of object is 5kg,
Fact #3: distance along inclined is L=5m.

(a) The initial kinetic energy is:

$$K_i = \frac{1}{2}mv^2 = \frac{1}{2}(5\text{kg})(10\,{}^{m}/_{s})^2 = 250\text{J}$$

By taking the ground as a reference, the final gravitational potential energy is given by

$$U_f = mgh$$

The height is calculated by using

$$h = L\sin\theta = 5\text{m}\sin 53 = 4.0\text{m}$$

Thus, the final potential energy is:

$$U_f = mgh = (5\text{kg})\left(10\,{}^{m}/_{s^2}\right)(4\text{m}) = 200\text{J}$$

Since the initial kinetic energy is greater than the final potential energy, then it means kinetic friction is present between the block and the inclined.

(b) The difference between the final potential and kinetic energies is the work done by the kinetic friction, i.e.

$$W_f = U_f - K_i = -50J$$

7. A block of mass 50 kg moving with a velocity of 20 m/s, on a horizontal frictionless surface, is brought to rest by an external force in 10 seconds. Calculate (a) the work done to completely stop the block, (b) the magnitude of applied force, and (c) the total distance it travelled before coming to stop.

Solution:

(a) Since the surface is frictionless, the only force that is doing work is the applied force. Thus, the work done by the external force is equivalent to the net work done on the object, which is the same as the change in the kinetic energy of the object.

Fact #1: mass of block is 50kg

Fact #2: initial velocity of block is 20m/s,

Fact #3: final velocity is 0,

Fact #4: time is 10s

Thus the work done is

$$W = \frac{1}{2}mv^2 - \frac{1}{2}mv_0^2 = 0 - \frac{1}{2}(50kg)(20\,{}^{m}/_{s})^2 = -10000J$$

(b) The applied force is obtained by applying Newton's 2nd law, i.e.

$$\vec{F} = m\vec{a}$$

This means I have to find the acceleration of the block,

$$a = \frac{v - v_0}{t} = \frac{0 - 20\,{}^{m}/_{s}}{10s} = -2\,{}^{m}/_{s^2}$$

Thus the magnitude of the applied force is

$$|\vec{F}| = F = (50kg)\left(2\,{}^{m}/_{s^2}\right) = 100N$$

(c) I have alternate approaches to find the total distance travelled by the block. One approach is obtained by calculating the work done, i.e.

$$w = Fd = 1000J \Rightarrow d = \frac{-10000J}{-100N} = 100m$$

8. A 200 kg aid package is dropped by a low flying aircraft from an altitude of 100m above the ground. If the aircraft was flying horizontally with a speed of 50m/s, what is the kinetic energy of the package just before it hits the ground?

Solution:
Fact #1: initial velocity of package is $v_0=50m/s$,
Fact #2: height above ground is $h=100m$,
Fact #3: acceleration due to gravity is approximately $10m/s^2$.
By applying the work-kinetic energy theorem, which states that the amount of work done on an object is equal to the change in its kinetic energy, I can easily solve this problem i.e.

$$W = \Delta K = K_f - K_i \qquad (1)$$

Solving for the final kinetic energy yields,

$$K_f = W + K_i \qquad (2)$$

Here work is done on the package by the force of gravity, i.e.

$$\mathrm{W = mgh = 200kg\left(10\,{}^{m}/_{s^2}\right)(100m) = 2 \times 10^5\ J}$$

Substituting the newly calculated value for the work into Equation 2 yields the final kinetic energy of the package as impacts the ground is

$$K_f = 2 \times 10^5\ \mathrm{J} + \frac{1}{2}(200kg)(50\,{}^{m}/_{s})^2 = 450{,}000J$$

9. A 1000kg-racing car is travelling at a top speed of 180km/h. If the driver applies the brakes to stop the car, which scenario results in the brakes doing more work: (a) slowing down from a speed of 180km/h to 90 km/h or (b) slowing down from 90km/h to a stop? Justify your answer.

Solution:
Fact #1: mass of car is $m=1000kg$,
Fact #2: initial speed for the first stage is $v_1=180$ km/h=50 m/s,
Fact #3: final speed for the first stage is $v_2=90$ km/h=25m/s,
Fact #4: initial speed for the second stage is $v_3=90$ km/h=25 m/s,
Fact #5: final speed for the second stage is $v_4=0$ m/s.

Work is done by the brakes to change the kinetic energy of the car. This means I have to calculate the change in the kinetic energies for the two scenarios and then compare the two results. Calculating the change in the kinetic energy for each case yields,

$$\mathrm{\Delta K_1 = \frac{1}{2}(1000kg)(50\,{}^{m}/_{s})^2 - \frac{1}{2}(1000kg)(25\,{}^{m}/_{s})^2 = 937{,}500J}$$

$$\mathrm{\Delta K_2 = \frac{1}{2}(1000kg)(25\,{}^{m}/_{s})^2 - 0 = 312{,}500J}$$

Since the change in kinetic energy for the first stage is larger than that during the second stage, the work done by the brakes is greater when the car's speed is reduced from 180 km/h to 90km/h than the work done in reducing the speed from 90km/h to a stop.

10. A 100 kg box that is initially at rest is pulled 10 m along a rough, horizontal floor with a constant applied horizontal force of 300N acting as shown in the figure below.

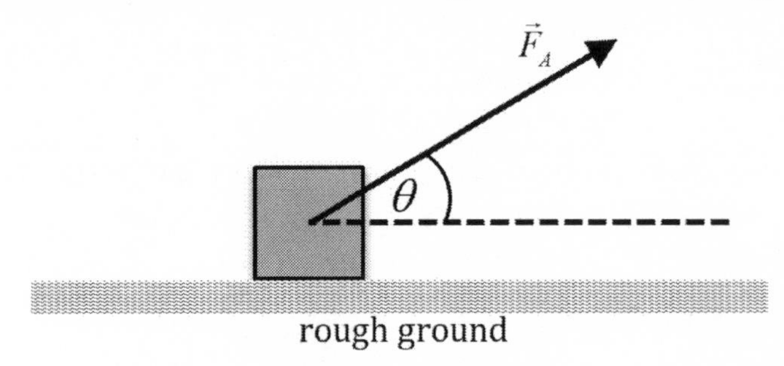

If the coefficient of friction between box and floor is 0.15 and θ=45^0, (a) find the work done by the applied force, (b) find the work done by the frictional force, (c) find the change in the kinetic energy of the box, and (d) find the final speed of the box.

Solution:
Fact #1: distance is d=10m,
Fact #2: magnitude of applied force is F_A=300N,
Fact #3: mass of box is m=100kg.
Fact #4: coefficient of kinetic friction is μ_k=0.15,
Fact #5: acceleration due to gravity is g≅10m/s^2.

(a) and (b). Since all forces acting on the box are constants, the work done by each force is obtained by multiplying the component of each force in the direction of the displacement of the box by the magnitude of the displacement (distance). Let W_A and W_f be the work done by the applied and frictional forces respectively,

$$W_A = (300\text{N})\cos 45\,(10\text{m}) = 2121.32\text{J} \qquad (1)$$

$$W_f = f(10\text{m})\cos\langle f, d\rangle \qquad (2)$$

Since the frictional force is not yet known, it is time to take a detour to find it. This is done by first drawing the free-body diagram showing all the forces acting on the box.

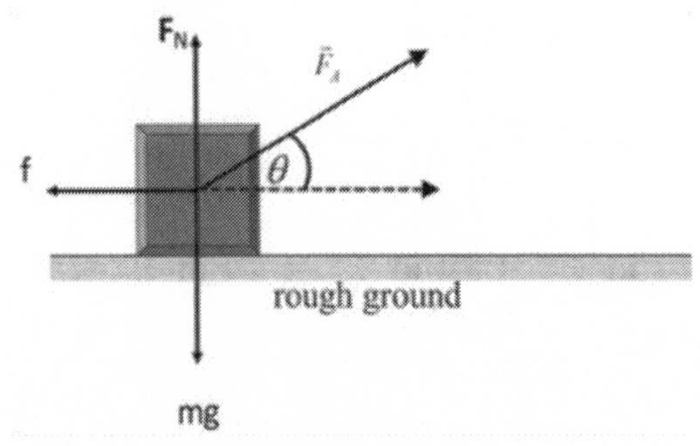

The vertical components of the forces must balance with each other, i.e.

$$F_N + F_A \sin 45 - mg = 0 \qquad (3)$$

Solving for the normal force yields,

$$F_N = mg - F_A \sin 45 = (100\text{kg})\left(10\,{}^{m}/_{s^2}\right) - 300\text{N} \sin 45 = 787.87\text{N}$$

Now I am in a position to find the magnitude of the frictional force by multiplying the normal force by the coefficient of friction, i.e.

$$f = \mu_k F_N = (0.15)(787.857\text{N}) = 118.18\text{N} \qquad (4)$$

The work done by the frictional force is obtained by substituting Eq. 4 into Eq. 2. The frictional force is opposite to the direction of the displacement, which means that the angle between the direction of the frictional force and displacement is 180^0. Thus the work done by the frictional force is

$$W_f = f(10\text{m}) \cos\langle f, d\rangle = (118.18\text{N})10\text{m} \cos 180 = -1181.8\text{J}$$

(c) The change in the kinetic energy of the box is the sum of the total work done by all the forces. The work done by the normal force and the weight of the box are both zero, because they are acting perpendicular to the direction of the displacement of the box. Thus,

$$\Delta K = W_A + W_f = 2121.32\text{J} - 1181.8\text{J} = 939.52\text{J}$$

(d) The final speed of the box is obtained by applying the definition of kinetic energy, i.e.

$$\Delta K = K_f - K_i \Rightarrow \Delta K = K_f = \frac{1}{2} m v_f^2 \qquad (5)$$

Solving for the final speed in Eq. 5 yields,

$$v_f = \sqrt{2\frac{\Delta K}{m}} = \sqrt{\frac{2(939.52\text{J})}{100kg}} = 4.33\,{}^{m}/_{s}$$

11. If a 50N force stretches a spring obeying Hooke's law by 5cm find (a) the work done in stretching the spring and (b) the work needed to stretch the spring by an additional 10cm.

Solution:
Fact #1: the force is F=50N,
Fact #2: amount of stretch is x =5cm=0.05m,
Fact #3: spring constant is k=unknown,
Fact #4: initial work done is W_1= unknown,
Fact #5: final work done is W_2=unknown.

(a) According to Hooke's law, the magnitude of the force exerted by a stretched spring is;

$$F = kx \qquad (1)$$

The work done in stretching a spring is given by

$$W_1 = \frac{1}{2}kx^2 \qquad (2)$$

In order to use Equation (2) to calculate the work done, I need to know the spring constant. The spring constant can be calculated from Eq. (1), i.e.

$$\mathrm{k} = \frac{\mathrm{F}}{\mathrm{x}} = \frac{50\mathrm{N}}{0.05\mathrm{m}} = 1000\,{}^{\mathrm{N}}/_{\mathrm{m}}$$

Thus, the work done in stretching the spring by 5cm is,

$$\mathrm{W_1} = \frac{1}{2}\left(1000\frac{\mathrm{N}}{\mathrm{m}}\right)(0.05\mathrm{m})^2 = 1.25\mathrm{J}$$

(b) The work done to stretch the spring by an additional 10 cm is calculated as follows. Let W be the work done to stretch the spring by 15cm, where

$$\mathrm{W} = \frac{1}{2}\left(1000\frac{\mathrm{N}}{\mathrm{m}}\right)(0.15\mathrm{m})^2 = 11.25\mathrm{J}$$

Thus, the work done to stretch it by the additional length is the difference between W and W_1.

$$\mathrm{W_2 = W - W_1 \Rightarrow W_2 = 11.25J - 1.25J = 10J}$$

Chapter Seven: Potential and Mechanical Energy

1. In the figure below a block of mass 2 kg is stationary at point A. There is no friction between the block and the ramp. There is a spring to the right having a spring constant of 5000 N/m.

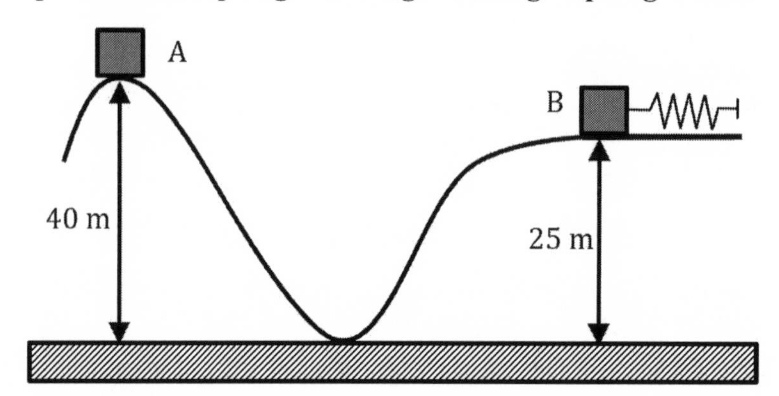

(a) What are the block's kinetic and potential energies at point A, (b) what are the block's kinetic and potential energies at position B, (c) what is the velocity of the block at point B, and (d) how far does the spring compress when the block stops?

Solution
Fact #1: height of block is h_A=40m,
Fact #2: mass of block is 2.0kg
Fact #3: speed block at A, v_A=0,

I know that the gravitational potential energy depends on the vertical position of the object and its kinetic energy is a function of its motion. Now taking the reference to be at ground level, the potential energy at A is given by

$$U_A = mgh_A = (2\text{kg})\left(10\,{}^{\text{m}}/_{\text{s}^2}\right)(40\text{m}) = 800\text{J}$$

The kinetic energy of the block at A is zero since its velocity there is zero.

$$K_A = 0$$

(b) & (c). The potential energy and the kinetic energy at B are given by

$$U_B = mgh_B = (2\text{kg})\left(10\,{}^{\text{m}}/_{\text{s}^2}\right)(25\text{m}) = 500\text{J}$$

$$K_B = \left({}^{1}/_{2}\right)mv_B^2 \qquad (1)$$

Since I do not know the speed of the block at B, I have to take a detour at this point. Because there is no friction between the block and the surface, the mechanical energy is conserved, i.e.

$$E_A = E_B$$

Where

$$E_A = K_A + U_A = 800\text{J} \qquad (2)$$

$$E_B = K_B + U_B = \left({}^{1}/_{2}\right)(2\text{kg})v_B^2 + 500\text{J}$$

Equating E_A and E_B and solving for v_B yields,

$$\left({}^{1}/_{2}\right)(2\text{kg})v_B^2 + 500\text{J} = 800\text{J}$$

Thus I get

$$v_B^2 = 300\text{J/kg}$$

Substituting this value of v_B into Eq. (1) gives

$$K_B = \left({}^{1}/_{2}\right)mv_B^2 = \left(\frac{1}{2}\right)(2\text{kg})(300\text{J}) = 300\text{J}$$

(d) Since there is no friction, all the mechanical energy of the block will be converted into elastic potential energy of the spring, i.e.

$$U_p = \left(\frac{1}{2}\right)kx^2 \qquad (3)$$

To find x I have equate Eqs. (2) and (3),

$$\left(\frac{1}{2}\right)kx^2 = 800\text{J} \Rightarrow x = 0.57\text{m}$$

Therefore, I have found that the initial 800J of mechanical energy at position A converted into compressing the spring by 0.57 m.

2. A 1.0 kg object is released from rest at a height of 12.0 m on a curved frictionless ramp. At the foot of the ramp is a spring of force constant 5000 N/m. If the object slides down the ramp and into the spring, compressing it a distance x before coming momentarily to rest, (a) find the velocity of the block at the bottom of the ramp and (b) find the amount of compression.

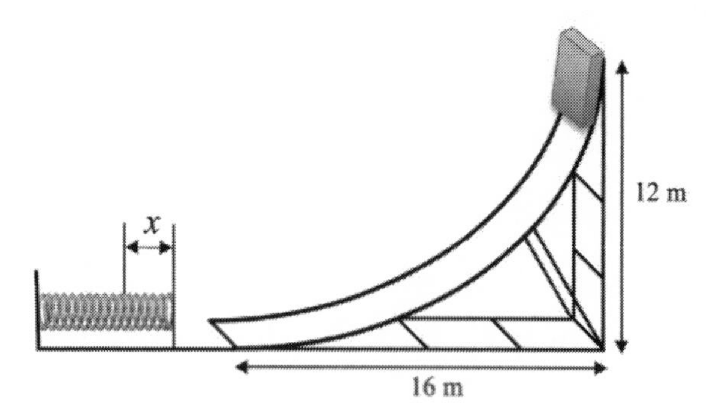

Solution:
If the reference for potential energy is zero at the bottom of the ramp, then the block at the top of the ramp has a potential energy given by

$$U_{top} = mgh = (1.0\text{kg})\left(10\,{}^{m}/_{s^2}\right)(12\text{m}) = 120\text{J}$$

At the bottom of the ramp the energy becomes completely kinetic until it hits the spring. When the block stops, by compressing the spring, it has transferred all its kinetic energy into elastic potential energy, i.e.

$$U_p = \frac{1}{2}kx^2 = 240\text{J} \Rightarrow x = \sqrt{\frac{240\text{J}}{5000\,{}^{N}/_{m}}} = 0.22\text{m}$$

3. In the figure below a skier of mass 60kg is stationary at the top of a hill of height H=50m. He starts from rest and slides down the hill where there is a ramp inclined at an angle of θ=37⁰.

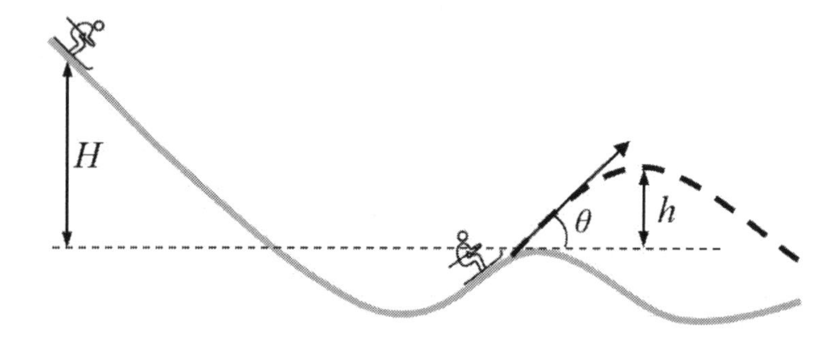

(a) What are the skier's kinetic and potential energy at the top of the hill, (b) what are the skier's kinetic and potential energy at the end of the ramp, (c) what is the velocity of the skier at the end of the ramp, and (d) how high does the skier go into the air after the ramp?

Solution:

Fact #1: mass of skier is m=60kg,

Fact #2: height of initial part of the hill is H=50m,

Fact #3: acceleration due to gravity is $g \cong 10 m/s^2$,

Fact #4: initial velocity of skier is $v_0=0$

Since the skier is not moving, his kinetic energy at the top of the hill is zero. By taking my reference at the bottom of the hill, the gravitational potential energy of the skier at the top of the hill is obtained by using

$$U_{top} = mgH = (60kg)\left(10\,{}^{m}/_{s^2}\right)(50m) = 30{,}000J$$

(a) At the bottom of the hill the potential energy of the skier is zero. Since the hill is frictionless there is no loss of energy. Thus the kinetic energy at the bottom of the hill is equal to his gravitational potential energy when he was at the top of the hill, i.e.

$$K_{bot} = U_{top} = 30{,}000J$$

(b) The velocity of the skier at the bottom of the hill is obtained form the kinetic energy, i.e.

$$K_{bot} = 30{,}000J = \frac{1}{2}mv_{bot}^2 \Rightarrow v_{bot} = 31.62\,{}^{m}/_{s}$$

(c) The skier becomes a projectile after he leaves the ramp. The vertical component of the velocity as he leaves the ramp is responsible for how high he can rise. The initial vertical component of the velocity is obtained by

$$v_{0y} = v_0 \sin\theta = (31.62\,{}^{m}/_{s}) \sin 37 = 18.97\,{}^{m}/_{s}$$

I can use conservation of energy to calculate how high the skier can go up. The initial kinetic as he leaves the ramp is

$$K_i = \frac{1}{2}m(v_{0y})^2 \qquad (1)$$

When he reaches the maximum height, all the initial kinetic energy is converted into gravitational potential energy, i.e.

$$U_f = mgh \qquad (2)$$

Equating Eqs. 1 and 2, and then solving for the height yields,

$$mgh = \frac{1}{2}m(v_{0y})^2 \Rightarrow h = \frac{(0.5)(18.97\,{}^{m}/_{s})^2}{10\,{}^{m}/_{s^2}} = 18m$$

4. A 3kg block is dropped from a height of 500 cm onto a spring having a spring constant of 5000 N/m.

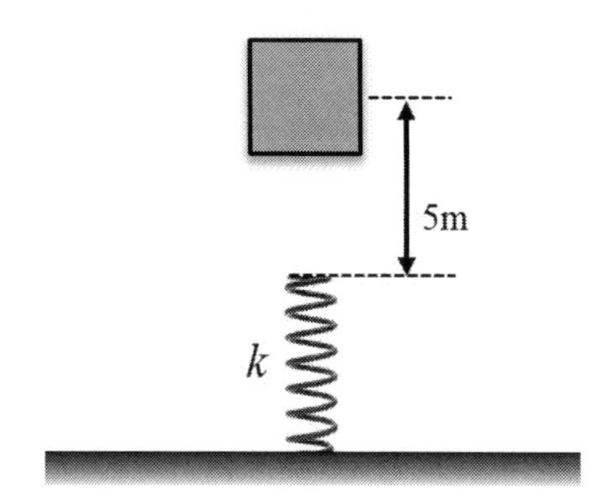

(a) Find the speed of the block when the compression of the spring is 10 cm, (b) How much farther must the spring be compressed for the block to momentarily come to rest?

Solution:
Fact #1: mass of block is m=3kg,
Fact #2: spring constant is k=5000N/m,
Fact #3: distance between the block and the spring is h=5m,
Fact #4: acceleration due to gravity is $g \cong 10 m/s^2$,
Fact #5: speed of the block when the spring is compressed is v=unknown.

(a) This problem is solved by applying the conservation of mechanical energy. Taking the initial position of the block as the zero of the gravitational potential energy, and because the velocity is zero, the initial mechanical energy is also zero, i.e.

$$E_i = 0 \qquad (1)$$

When the spring is compressed by 10cm, the block has fallen by 5.1m from its initial position. Since the block is still moving as it compresses the spring by 10cm, it possesses kinetic energy as well as gravitational and elastic potential energies.

$$E_f = \frac{1}{2}mv^2 - mg(5.1m) + \frac{1}{2}k(0.1m)^2 \qquad (2)$$

Equating Eqs. (1) and (2) and solving for the speed yields,

$$\frac{1}{2}mv^2 - mg(5.1m) + \frac{1}{2}k(0.1m)^2 = 0 \Rightarrow v = \sqrt{\frac{2(153J) - 50J}{3kg}} = 9.24\, {}^{m}/_{s}$$

(a) When the block momentarily comes to rest, all the mechanical energy is in the form of gravitational and elastic potential energy, i.e.

$$E_f' = -mg(5 + x) + \frac{1}{2}kx^2 \qquad (3)$$

Once again applying conservation of mechanical energy yields,

$$-mg(5 + x) + \frac{1}{2}kx^2 = 0 \Rightarrow 250x^2 - 3x - 15 = 0 \qquad (4)$$

Solving for x gives

$$x = 0.25m \quad or\ x = -0.24m$$

The negative value of x is discarded. The additional amount the spring must be compressed after the initial compression is

$$x' = 0.25\text{m} - 0.1\text{m} = 0.15\text{m}$$

5. A bullet that has been fired straight down from the top of a building hits the ground with a kinetic energy of 100J. If another bullet is fired straight up with the same muzzle velocity with what kinetic energy will the second bullet hit the ground? Justify your answer.

Solution:

The second bullet goes up until its velocity becomes zero and reverses course and falls straight back. On its way to the ground from its maximum height, the speed of the bullet begins to increase and regains all the speed it had lost on its way up. This is like a zero-sum game. The fact that the bullet was fired straight up does not make any difference in the kinetic energy of the bullet from the bullet that was fired straight down. Thus, the second bullet also will hit the ground with a kinetic energy of 100J.

6. A long spring having a spring constant of 50 N/m is compressed by 0.15m from its equilibrium position. How much work must be done to compress it an additional 0.1 m?

Solution:

Fact #1: spring constant is k=50N/m,

Fact #2: amount of initial compression is x_1=0.15m,

Fact #3: Work done during initial compression is W_1=unknown,

Fact #4: Work done during second stage of compression is W_2=unknown,

Fact #5: Total work done during both stages of compression of the spring is W=unknown,

The work done in the initial stage is given by

$$W_1 = \frac{1}{2}kx_1^2 = 0.5(50\,{}^{\text{N}}/_{\text{m}})(0.15\text{m})^2 = 0.5625\text{J} \qquad (1)$$

The total work done is obtained as follows,

$$W = \frac{1}{2}kx^2 = 0.5(50\,{}^{\text{N}}/_{\text{m}})(0.25\text{m})^2 = 1.5625\text{J} \qquad (2)$$

Thus the work done to compress the spring by an additional 0.1m is

$$W_2 = W - W_1 = 1.5625\text{J} - 0.5625\text{J} = 1.0\text{J}$$

6. **A 2 kg projectile is fired with a muzzle velocity of 500m/s at an angle of 37^0 above the horizontal in calm air. Find the kinetic and gravitational potential energies of the projectile when it reaches its highest point.**

Solution:
Fact #1: mass of projectile is m=2kg,
Fact #2: initial velocity is v_0=500m/s,
Fact #3: angle of projectile is 37^0,
Fact #4: acceleration due to gravity is g≅$10m/s^2$ (downwards),
Fact #5: maximum height of projectile is y=unknown,
Fact #6: initial vertical component of the velocity is v_{0y}=(500m/s)sin37=300m/s,
Fact #7: initial horizontal component of the velocity is v_{0x}=(500m/s)cos37=400m/s,
Fact #8: final vertical component of velocity is 0.

A common mistake that students make in this problem is assuming the velocity of the projectile at its highest point to be zero. Although the vertical component is zero, the horizontal component is never zero at all; hence, the projectile is never stationary at any location in the air. As usual, let me list the facts of the problem. Since the air is calm, the horizontal component of the projectile is constant throughout its flight. This means that when the projectile reaches its maximum height, it is still moving horizontally with a speed of 400m/s because the vertical component has become zero. Hence, the kinetic energy is solely due to the horizontal component of the velocity. The kinetic energy of the projectile is therefore,

$$K = \frac{1}{2}mv_{ox}^2 = (0.5)(2kg)(400m/s)^2 = 1600J$$

In order to find the gravitational potential energy, I have to find the maximum height the projectile can reach. The maximum height is obtained by using one of the equations of motions. Since the acceleration, initial and final vertical component of the velocity are known, the following equation can be used to find the maximum height, i.e.

$$y = \frac{v_y^2 - v_{0y}^2}{2g} = \frac{0 - (300\,{}^{m}/_{s})^2}{2(-10\,{}^{m}/_{s^2})} = 4500m$$

Thus the gravitational potential energy is

$$U_g = mgh = (2kg)\left(10\,{}^{m}/_{s^2}\right)(4500J) = 90{,}000J$$

8. A simple pendulum consists of a 50g bob suspended from a frictionless pivot by using a string of length 1.0m as shown in the figure below.

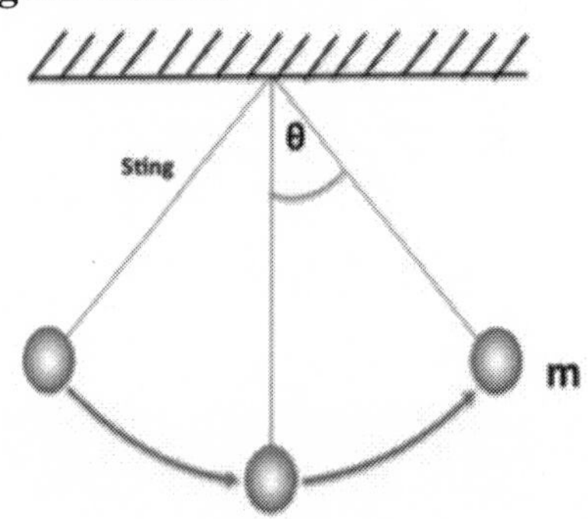

If the mass is pulled by a small angle of 20^0, (a) what is the speed of the bob at its lowest point, and (b) what is its maximum gravitational potential energy when it swings all the way to the maximum on the opposite side?

Solution:
As the pendulum swings, back and forth its mechanical energy is continuously changing between kinetic and gravitational energies. Let h be the maximum height that the mass rises to from the lowest point. The height is obtained by using trigonometric relations, i.e.

$$h = \ell - \ell \cos \theta = 1\text{m} - 1\text{m}\cos 20 = 0.06\text{m}$$

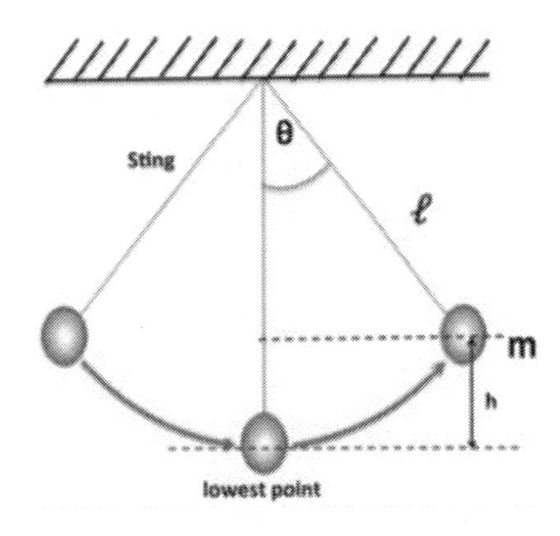

If I take the gravitational potential energy to be zero at the lowest point, then the gravitational potential energy at the highest point is;

$$U = mgh = (0.02\text{kg})\left(9.8\,{}^{\text{m}}/_{\text{s}^2}\right)(0.06\text{m}) = 0.012\text{J} = 12\text{mJ}$$

At the lowest point, all the gravitational potential energy has become kinetic energy, i.e.

$$K = \frac{1}{2}\text{mv}^2 = 0.012\text{J}$$

Thus, the maximum speed of the bob is

$$v = \sqrt{\frac{2(0.012\text{J})}{0.02kg}} = 1.1\,{}^{m}/_{s}$$

9. A 100g object has been shot straight up into the air with an initial velocity of 50m/s. If the object reaches a maximum height of 100m, (a) find the gravitational potential energy of the object at its highest point, and (b) find the amount of energy lost due to air resistance.

Solution:
As usual let me list the facts.
Fact #1: mass of object is m=0.1kg,
Fact #2: initial velocity of object is v_0=50m/s,
Fact #3: maximum height reached is h=100m,
Fact #4: acceleration due to gravity is g=10m/s^2.

a. The gravitational potential energy is obtained by using,

$$U_g = mgh = (0.1\text{kg})\left(10\,{}^{m}/_{s^2}\right)(100\text{m}) = 100\text{J}$$

If there were no air resistance, the object would have risen higher than 100m because its initial kinetic energy is greater than 100J, i.e.

$$K_0 = \frac{1}{2}mv_0^2 = \frac{1}{2}(0.1\text{kg})(50\,{}^{m}/_{s})^2 = 125\text{J}$$

b. The difference in the initial kinetic energy and the gravitational energy at the maximum is the amount of energy lost due to air resistance. Hence, the energy lost is:

$$E_{lost} = 125J - 100J = 25J$$

10. In the figure below, a spring gun (with spring constant *k=2500N/m*) is used to launch a block of mass 600g from a lower level to a higher level after having passed through an intermediate valley. In this problem, the track is frictionless.

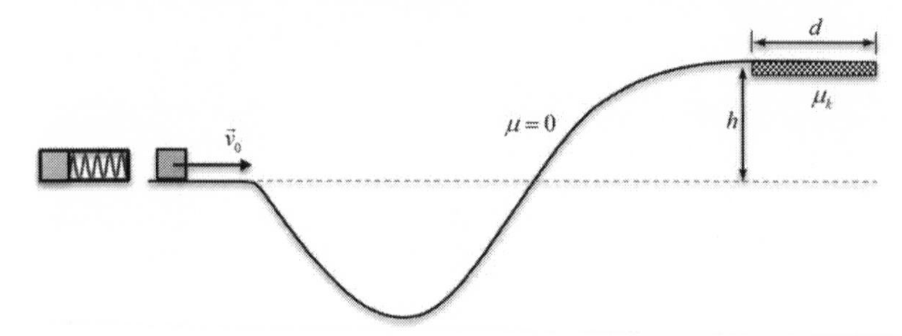

(a) How much must the spring be compressed so that the block will make it to the top of the hill on the right with a velocity of 5m/s? Here h=2.0m. (b) If the coefficient of kinetic friction is μ_k=0.3, how far does the block travel before it stops?

Solution:

(a) When the spring is compressed, it stores elastic potential energy, which later gets converted into kinetic energy. The elastic energy is calculated from

$$U = \frac{1}{2}kx^2 \qquad (1)$$

When the block makes it to the other side, the initial elastic energy is converted into kinetic and gravitational energy, i.e.

$$E = \frac{1}{2}mv^2 + mgh \qquad (2)$$

Equating Eqs. (1) and (2) yields

$$\frac{1}{2}kx^2 = \frac{1}{2}mv^2 + mgh \qquad (3)$$

Solving for x gives

$$x = \sqrt{\frac{(mv^2 + 2mgh)}{k}} = \sqrt{\frac{(.6kg(2\,{}^{m}/_{s})^2 + 2(0.6kg)\left(10\,{}^{m}/_{s^2}\right)(2m))}{2500\,{}^{N}/_{m}}} = 0.10m$$

(b) When the block stops after a distance of *d*, all the kinetic energy has been used to do work against friction, i.e.

$$W_f = f_k d = \frac{1}{2}mv^2 \qquad (4)$$

Here the frictional force is given by

$$f_k = \mu_k N = \mu_k mg = 0.3(0.6kg)\left(10\,{}^{m}/_{s^2}\right) = 1.8N \qquad (5)$$

Substituting the value obtained in Eq. (5) into Eq. (4) and solving for the distance *d*, yields,

$$d = \frac{mv^2}{f_k} = \frac{(0.6kg)(2\,{}^{m}/_{s})^2}{1.8N} = 1.33m$$

Additional Tidbits

a) How much elastic potential energy does a spring whose spring constant is 10N/m store when it is in its natural length (upstretched length)?

Answer To Tidbits

The amount of energy stored in an unstretched spring is zero.

Chapter Eight: Momentum and Related Topics

1. The figure below shows a force as a function of time acting on a 5.0 kg object. If the force acts for the duration shown, (a) find the impulse due to the force, and (b) find the final velocity of the particle if it is initially at rest.

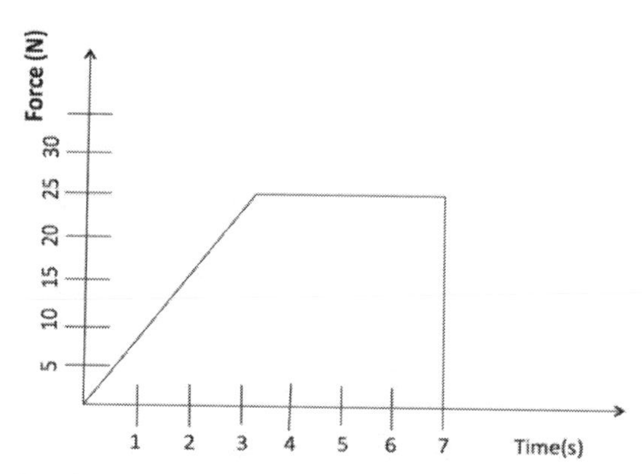

Solution:

(a) I know the area under the curve of force as a function of time gives the impulse generated by the force. Thus, all I have to do is find the area under the curve. Since the shape of the curve is not a single regular shape, I will divide it into two figures as shown in the figure below.

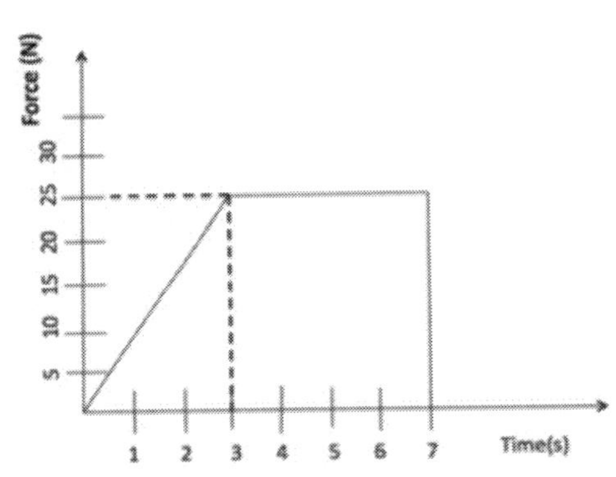

Now I have two regular shaped figures, i.e. a triangle and a rectangle. If I assign A_1 and A_2 to represent the area of the triangle and rectangle respectively then;

$$A_1 = \left(\frac{1}{2}\right)(3s)(25N) = 37.5Ns = 37.5\ ^{kgm}/_{s}$$

$$A_2 = (4s)(25N) = 100Ns = 100\ ^{kgm}/_{s}$$

Thus, the total area is the sum of A_1 and A_2, i.e.

$$A = A_1 + A_2 = 37.5\ ^{kgm}/_{s} + 100\ ^{kgm}/_{s} = 137.5\ ^{kgm}/_{s}$$

Hence, the impulse due to the force is 137.5 kgm/s

(b) To find the final velocity, I proceed as follows. I know impulse is the change in momentum. This means I have to find the initial and final momenta and take the difference. Let P_i and P_f be the initial and final momentum respectively,

$$P_i = mv_i = 0$$

$$P_f = mv_f = (10kg)v_f$$

Now finding the change in the momentum gives

$$\Delta P = P_f - P_i = (10kg)v_f - 0 = 137.5\ ^{kgm}/_{s}$$

Solving for the final velocity yields,

$$v_f = 13.5\ ^{m}/_{s}$$

2. A 10 kg object moving with 100 kgm/s momentum collides with a 15kg object moving with a momentum of 300 kgm/s in a direction of 37 degrees from the initial direction of the 10 kg object, as shown below. (a) If they remain stuck together after the collision what concept will you apply to find the speed of the two objects after the collision? (b) Find the speed of the two objects after the collision, and (c) is the collision elastic or inelastic? Justify your answer.

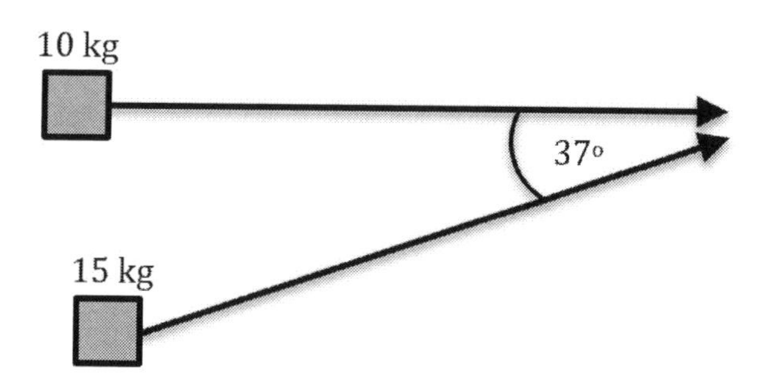

Solution:

(a) Since the collision is the result of the interaction of the two particles without the involvement of an external force, the sum of the momentum of each object must be conserved. Thus, this is a momentum conserving interaction. Hence I will apply the conservation of momentum to find the speed of the objects after the collision.

(b) Here I have to remember that momentum is a vector quantity. This means I have to consider the direction of the motions of the interacting particles. For convenience, let the line of motion of the 10kg object be horizontal. Thus, I have horizontal and vertical directions. If P_{1i} and P_{2i} are the momentum of each object before the collision, then I can use an analytical approach to find the total momentum before collision.

Momentum	horizontal-component	vertical component
P_{1i}	100 kgm/s	0
P_{2i}	(300kgm/s) cos37=240kgm/s	(300kgm/s)sin37=180kgm/s

By using **unit** vectors, I can write the initial momentum as

$$\vec{P}_i = \vec{P}_{1i} + \vec{P}_{2i} = \left(100\, {}^{kgm}/_{s} + 240\, {}^{kgm}/_{s}\right)\vec{\imath} + (180)\, {}^{kgm}/_{s})\vec{\jmath} \qquad (1)$$

Let P_f represent the final momentum of the two objects after the collision,

$$\vec{P}_f = m_1\vec{v}_{1f} + m_2\vec{v}_{2f}$$

Since the two objects are locked together after the collision then they have one common velocity, i.e.

$$\vec{v}_{1f} = \vec{v}_{2f} = \vec{v}_f$$

Thus, the final momentum can be modified as follows.

$$\vec{P}_f = m_1\vec{v}_{1f} + m_2\vec{v}_{2f} = (m_1 + m_2)\vec{v}_f = (25kg)\vec{v}_f \qquad (2)$$

Here equating Equations 1 and 2 and solving for the final velocity yields,

$$\vec{v}_f = \frac{\left(340\ ^{kgm}/_{s}\right)\vec{\imath} + \left(180\ ^{kgm}/_{s}\right)\vec{\jmath}}{25kg} = (13.6\ ^{m}/_{s})\vec{\imath} + (7.2\ ^{m}/_{s})\vec{\jmath}$$

The speed, which is the magnitude of the final velocity, is obtained like any other vector, i.e.

$$|\vec{v}_f| = v_f = \sqrt{(13.6\ ^{m}/_{s})^2 + (7.2\ ^{m}/_{s})^2} = 15.39\ ^{m}/_{s}$$

Thus, the speed of the two objects locked together is 15.39m/s.

(c) In order to find out whether the collision is elastic or inelastic, I have to obtain the total kinetic energies before and after collision and compare the two results. Since I know the final speed, it is easy to compute the final kinetic energy, i.e.

$$K_f = \frac{1}{2}(m_1 + m_2){v_f}^2 = \frac{1}{2}(25kg)(15.39\ ^{m}/_{s})^2 = 2960.65J$$

To find the initial kinetic energy, I have to find the speed of the two objects before the collision. The speed of the first object was 10m/s. However, for the second object, I have to obtain first its velocity from the initial momentum, i.e.

$$\vec{P}_{2i} = \left(240\ ^{kgm}/_{s}\right)\vec{\imath} + (180)\ ^{kgm}/_{s}\vec{\jmath} = (15kg)\vec{v}_{2i}$$

Thus, the initial velocity of the second object was

$$\vec{v}_{2i} = (16\ ^{m}/_{s})\vec{\imath} + (12\ ^{m}/_{s})\vec{\jmath}$$

Hence the initial speed of the second object was

$$|\vec{v}_{2i}| = v_{2i} = \sqrt{(16\ ^{m}/_{s})^2 + (12\ ^{m}/_{s})^2} = 20m/s$$

Thus, the total initial kinetic energy is

$$K_i = \frac{1}{2}(10kg)(10\ ^{m}/_{s})^2 + \frac{1}{2}(15kg)(20\ ^{m}/_{s})^2 = 3500J$$

Because the final kinetic energy is not equal to the initial kinetic energy, the collision was inelastic.

3. In order to determine the velocity of a bullet, Jack is conducting an experiment. In this experiment, he fires his bullet into a suspended wooden block with a velocity of v_0 as shown below. The bullet imbeds itself into the wooden block and the two swing to a height of *h*, which Jack is able to measure. He then obtains the masses of the bullet and wooden block by using a scale. If the mass of the bullet (m_b) and mass of the wooden block (M) were 0.005kg and 2.0 kg respectively and the height is 10cm, find the speed of the bullet.

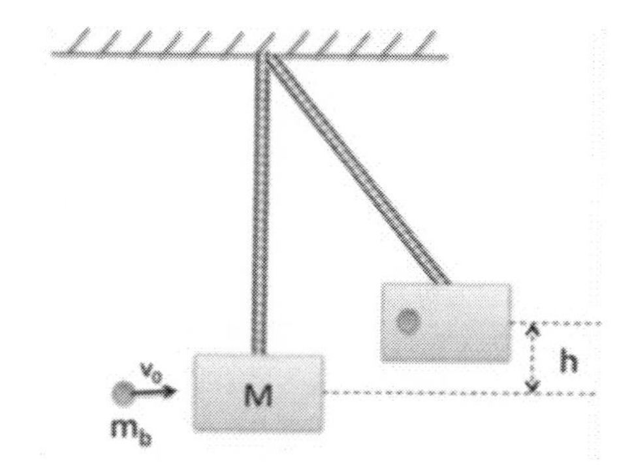

Solution:
Fact #1: initial speed of bullet is v_0=unknown,
Fact #2: mass of the bullet is m_b=0.005kg,
Fact #3: mass of wooden block is M=2.0kg,
Fact #4: initial speed of block is V_0=0,
Fact #5: height is h=10cm,
Fact #6: velocity of the bullet and block after collision is V_f=unknown.

This problem involves the collision of the bullet and the wooden block without any external force. Thus, I can find the speed of the bullet by applying conservation of momentum. Now I can write the initial and final momenta as follows.

$$P_i = m_b v_0$$
$$P_f = (m_b + M)V_f$$

Applying the conservation of momentum and solving for the velocity of the bullet and wood block immediately after the collision yields,

$$V_f = \frac{m_b v_0}{m_b + M} \qquad (1)$$

Here, I see that I am running into a problem where I have two unknowns and one equation. This means, as Professor John has taught me, I have to take a detour. I see that the kinetic energy of the two colliding objects immediately after the collision is changing into gravitational potential energy. If K_0 and U_f are the respective kinetic and potential energies, then

$$K_0 = \frac{1}{2}(m_b + M)V_f^2 \qquad (2)$$

$$U_f = (m_b + M)gh \qquad (3)$$

Equating Eqs. (2) and (3) and then solving for V_f, gives.

$$V_f = \sqrt{2gh} = \sqrt{2\left(10\,{}^{m}/_{s^2}\right)(0.1m)} = 1.41\,{}^{m}/_{s} \qquad (4)$$

Now, using the newly obtained velocity of the bullet and block in Eq. (1) gives the result I have been working very hard to obtain, i.e.

$$v_0 = \frac{(m_b + M)}{m_b} V_f = \frac{2.005kg}{0.005kg}(1.41\,{}^{m}/_{s}) = 565.4\,{}^{m}/_{s}$$

Thus, the speed of the bullet is 565.4 m/s.

4. Two identical balls are on a frictionless, horizontal billiard tabletop. The first ball, which was moving at 20 m/s, collides elastically with the second ball that is stationary, as shown in the figure below. Since the collision was glancing, the balls scatter after the collision. If the first ball scatters along a path at 37^0 to its original direction, as shown in the figure, find the speed of each particle and the direction of motion of the second ball.

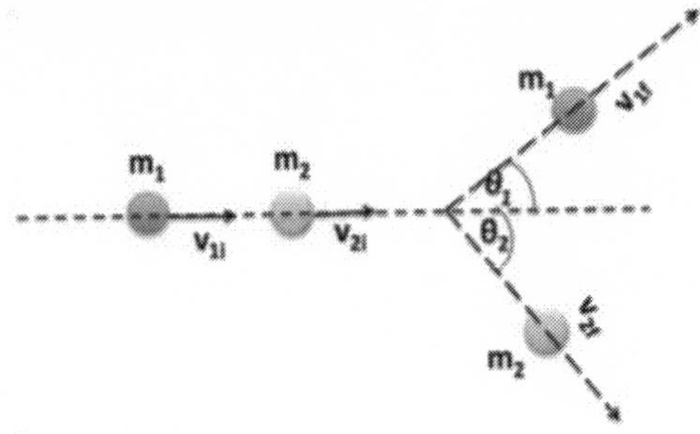

Solution:

Fact #1: initial velocity of m_1 is v_{1i}=20m/s,

Fact #2: initial velocity of m_2 is v_{2i}=0,

Fact #3: mass of first ball is m_1 =unknown,

Fact #4: mass of second ball is m_2 =unknown,

Fact #5: final velocity of first ball is v_{1f}=unknown,

Fact #6: final velocity of second ball is v_{2f}=unknown,

Fact #7: direction of first ball is $\theta_1=37^0$,

Fact #8: direction of second ball is θ_2=unknown.

I know that this is a conservation of momentum problem. It looks like I have several unknowns, but as Professor John has taught me, I am going to work my way and see how the problem unravels.

Momentum	horizontal-component	vertical-component
P_{1i}	$m_1v_{1i}=10m_1$	0
P_{2i}	0	0
P_{1f}	$m1v_{1f}\cos37=0.8m_1v_{1f}$	$m_1v_{1f}\sin37=0.6m_1v_{1f}$
P_{2f}	$m_2v_{2f}\cos\theta_2$	$-m_2v_{2f}\sin\theta_2$

Applying conservation of momentum for each component yields;

$$m_2v_{2f}\cos\theta_2 + 0.8m_1v_{if} = 10m_1 \qquad (1)$$

$$0.6m_1v_{1f} - m_2v_{2f}\sin\theta_2 = 0 \qquad (2)$$

Since the masses are equal, they cancel out from Eqs. 1 and 2 and upon rearranging and rewriting these equations I have

$$\cos\theta_2 = \frac{10 - (0.8v_{if})}{v_{2f}} \qquad (3)$$

$$\sin\theta_2 = \frac{0.6v_{1f}}{v_{2f}} \qquad (4)$$

Squaring Eqs. 3 and 4 and adding them gives

$$\cos^2\theta_2 + \sin^2\theta_2 = \left(\frac{10-(0.8v_{1f})}{v_{2f}}\right)^2 + \left(\frac{0.6v_{1f}}{v_{2f}}\right)^2 \qquad (5)$$

The left hand side of Eq. 5 is an identity element, i.e.

$$\cos^2\theta_2 + \sin^2\theta_2 = 1$$

Now I am ready to solve for the final speed of the second ball. Using Eqs 3, 4 and 5 gives

$$\left(\frac{10-(0.8v_{1f})}{v_{2f}}\right)^2 + \left(\frac{0.6v_{1f}}{v_{2f}}\right)^2 = 1 \qquad (6)$$

This can be modified into

$$100 - 16v_{1f} + v_{1f}^2 = v_{2f}^2 \qquad (7)$$

I still have two unknowns in Equation (7). I think I am stuck. I wish Professor John were here right now. Since he cannot be here, I have to remember what he said when I got stuck like now. He used to say, "did you play all your cards?" By that he meant did you use every bit of information given in the problem. This means, it is time to reread the problem carefully again and see what information was not used. I got it now. The collision was elastic, which means the kinetic energy is conserved, i.e.

$$\frac{1}{2}m_1v_{1f}^2 + \frac{1}{2}m_2v_{2f}^2 = \frac{1}{2}m_1v_{1i}^2$$

Simplifying this

$$v_{1f}^2 + v_{2f}^2 = 100 \qquad (8)$$

Hurray, I got the second equation that I needed. I can now use Eq. (8) to solve for one of the unknowns in terms of the other. Solving for v_{2f} yields

$$v_{2f}^2 = 100 - v_{1f}^2 \qquad (9)$$

Substituting this newly found expression of v_{2f} into Eq. (7) yields,

$$100 - 16v_{1f} + v_{1f}^2 = 100 - v_{1f}^2$$

Solving for v_{1f} gives

$$v_{1f} = 8\,{}^{m}/_{s} \quad \text{or} \quad v_{1f} = 0 \qquad (10)$$

The speed of the second ball is obtained by using this value of v_{1f} in Eq. 9, i.e.

$$v_{2f}^2 = 100 - 64 = 36 \quad \Rightarrow \quad v_{2f} = 6\,{}^{m}/_{s}$$

To find the value of θ_2, one can use either Eq. 3 or 4. Choosing Eq. 4 and obtaining the inverse sine gives

$$\theta_2 = \sin^{-1}\frac{0.6v_{1f}}{v_{2f}} = \sin^{-1}\frac{4.8}{6} = 53^0$$

5. **In the figure below two metal plates of different materials are joined together. Both plates are square and have dimensions of 1m by 1m. Plate A has a mass of 20kg and plate B has a mass of 50kg. Find the coordinates of the center of mass of the two plates.**

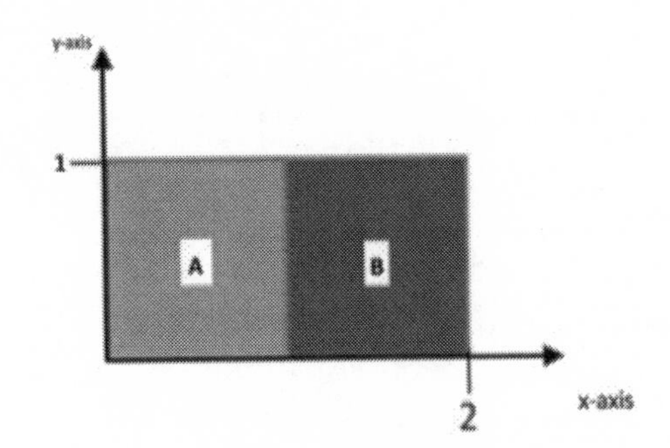

Solution:
Since the density of each block is uniform, by symmetry, each block's center of mass is at its geometric center.
If r_{cmA} and r_{cmB} represent the coordinates of the center of mass of objects A and B then,

$$r_{cmA} = (0.5m, 0.5m)$$
$$r_{cmB} = (1.5m, 0.5m)$$

Now, the objects can be considered as point particles. If x_{cm} and y_{cm} represent the coordinates of the center of mass of both "point particles" then we can write

$$x_{cm} = \frac{m_A x_1 + m_B x_2}{m_A + m_B} = \frac{(20kg)(0.5m) + (50kg)(1.5m)}{70kg} = 1.21m$$
$$y_{cm} = \frac{m_A y_1 + m_B y_2}{m_A + m_B} = \frac{(20kg)(0.5m) + (50kg)(0.5m)}{70kg} = 0.5m$$

Thus, the center of mass of the two objects is

$$(x_{cm}, y_{cm}) = (1.21m, 0.5m)$$

6. One of the metals plates in Problem #5 is now doubled as shown in the figure below. Plate A is still square with dimensions of 1m by 1m and has a mass of 20kg, whereas plate B now has dimensions of 1m by 2m and has a mass of 100kg. Find the coordinates of the center of mass.

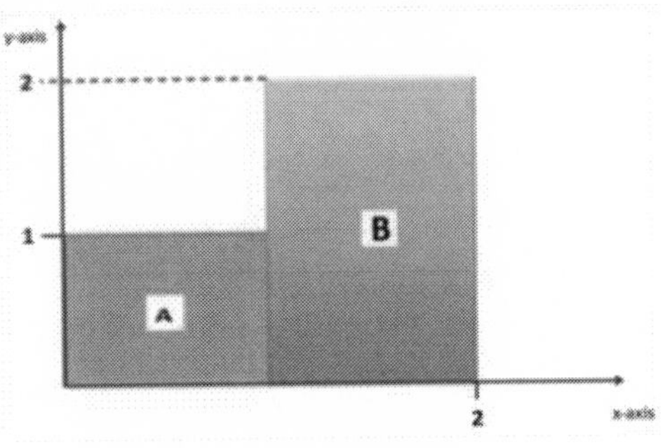

Solution:

As in Problem #5, the center of mass of each block is at its geometric center.

If r_{cmA} and r_{cmB} represent the coordinates of the center of mass of A and B then,

$$r_{cmA} = (0.5m, 0.5m)$$

$$r_{cmB} = (1.5m, 1.0m)$$

Now, the objects can be considered as point particles. If x_{cm} and y_{cm} represent the coordinates of the center of mass of the combination, then

$$x_{cm} = \frac{m_A x_1 + m_B x_2}{m_A + m_B} = \frac{(20kg)(0.5m) + (100kg)(1.5m)}{120kg} = 1.33m$$

$$y_{cm} = \frac{m_A y_1 + m_B y_2}{m_A + m_B} = \frac{(20kg)(0.5m) + (100kg)(1.0m)}{120kg} = 0.92m$$

Thus, the center of mass of the two objects is

$$(x_{cm}, y_{cm}) = (1.33m, 0.92m)$$

7. In the figure below is shown a can of soda of height D and radius R. Holes are drilled in the top and bottom of the can to let the soda drain. The density of the soda is ρ_s and the mass of the can is M_c.

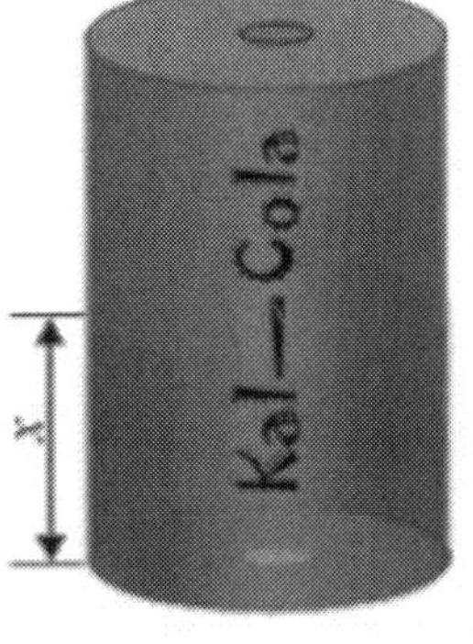

With *no* soda in the can, what is (a) the center of mass of just the CAN? (b) If the can is filled with soda to a level of *x*, what is the center of mass of just the SODA? (c) If the can is filled with soda to a level of *x*, what is the center of mass of the SODA and the CAN together?

Solution:

(a) Since the CAN is uniform, its mass is distributed evenly. Hence the center of mass is the same as its geometric center. Assuming the bottom left of the CAN is located at the origin (0,0), then the coordinates of the center of mass are given by

$$R_{cm} = \left(\frac{R}{2}, \frac{D}{2}\right)$$

(b) Once again since the soda is uniform, the center of mass coincides with geometric center of the soda, i.e.

$$r_{cm} = \left(\frac{R}{2}, \frac{x}{2}\right)$$

(c) Here the best approach is to treat the CAN and soda as two point objects. If I assign m_c and m_s to represent the masses of the CAN and soda respectively, then the center of mass of both is calculated as follows;

$$x_{cm} = \frac{m_c\left(\frac{R}{2}\right) + m_s\left(\frac{R}{2}\right)}{m_c + m_s} = \frac{R}{2}$$

$$z_{cm} = \frac{m_c\left(\frac{D}{2}\right) + m_s\left(\frac{x}{2}\right)}{m_c + m_s} = \frac{m_c D + m_s x}{2(m_c + m_s)}$$

Thus the coordinates of the center of mass of the CAN and soda system is given by

$$(x_{cm}, z_{cm}) = \left(\frac{R}{2}, \quad \frac{m_c D + m_s x}{2(m_c + m_s)}\right)$$

8. In the figure below, a bullet of mass *m=25g* is moving at an angle of $\theta=37^0$ above the horizontal and with an initial speed of v_{1i}. The bullet strikes a block of mass *M=3kg*, which is initially at rest v_{2i} = 0, and embeds itself into the block.

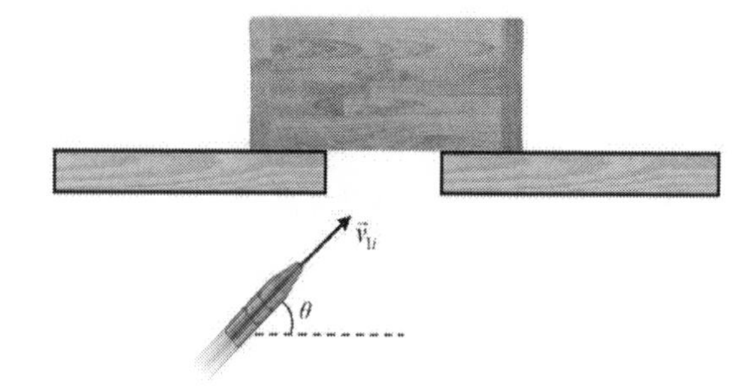

(a) If the block and bullet system land a distance of *5m* away, compute the initial velocity of the bullet just before impact. (b) Calculate the kinetic energy of the block-bullet system immediately after impact.

Solution:
Fact #1: initial velocity of bullet is v_{1i}=unknown,
Fact #2: initial velocity of the block is v_{2i}=0,
Fact #3: angle is $\theta=37^0$,
Fact #4: velocity of bullet and block immediately after the impact is v_f,
Fact #5: Horizontal distance is D=5m,
Fact #6: vertical distance is y=unknown,
Fact #7: time flight is t_f=unknown,

Since the bullet and block are the only interacting objects momentum is conserved. The initial momentum of the bullet and mass system is given by

$$p_i = mv_{1i} \qquad (1)$$

The final momentum of the bullet and mass system is

$$p_f = (m + M)v_f \qquad (2)$$

Applying conservation of momentum and solving for the initial velocity of the bullet yields,

$$v_{1i} = \frac{(m + M)v_f}{m} \qquad (3)$$

The bullet and block system describe projectile motion after impact. The horizontal and vertical components of the velocity after the collision are:

$$v_{fx} = v_f \cos\theta \quad and \quad v_{fy} = v_f \sin\theta \qquad (4)$$

The time of flight is determined by the vertical component, v_{fy}, i.e.

$$t_f = \frac{2v_f \sin\theta}{g}$$

The horizontal distance is the product of the horizontal component of the velocity and the time of flight, i.e.

$$D = v_f \cos\theta \frac{2v_f \sin\theta}{g} = \frac{2v_f^2 \sin\theta \cos\theta}{g} = 5m \qquad (5)$$

Solving for v_f yields

$$v_f = \sqrt{\frac{(5\text{m})g}{2\sin\theta\cos\theta}} = \sqrt{\frac{(5\text{m})\left(10\,\text{m}/\text{s}^2\right)}{2\sin 37\cos 37}} = 7.22\,{}^{m}/_{s} \qquad (6)$$

Substituting the result obtain in Eq. (6) into Eq. (3) gives the initial velocity of the bullet, i.e.

$$v_{1i} = \frac{(m + M)v_f}{m} = \frac{(0.025kg + 3kg)}{0.025kg}(7.22\,{}^{m}/_{s}) = 873.6\,{}^{m}/_{s}$$

9. A metallic ball of mass m=2.5kg strikes the ground with a speed of 10 m/s at an angle of $\theta=53^0$ with respect to the y-axis. The ball bounces off the ground with the same speed and angle, as shown in the figure. (a) Find the magnitude and direction of the impulse, and (b) if the ball is in contact with the ground for 0.2s, find the average force exerted on the ball by the ground.

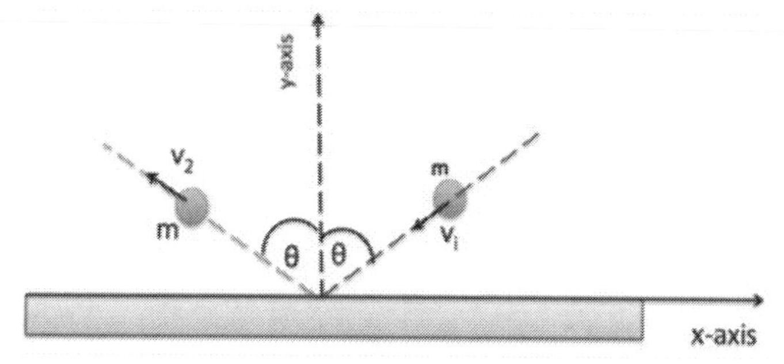

Solution:
Fact #1: mass is m=2.5kg,
Fact #2: initial speed is v_i=10m/s,
Fact #3: final speed is v_f=10m/s.
Fact #4: the initial and final speed are equal, i.e, $v_i=v_f=v$
(a) Impulse is defined as the change in momentum. Thus, I have to calculate the initial and final momentum of the ball. It is helpful to write the initial and final momentum as a sum of its components, i.e.

$$\vec{P}_i = -mv_i \sin\theta\, \vec{\imath} - mv_i \cos\theta\, \vec{\jmath}$$
$$\vec{P}_f = -mv_f \sin\theta\, \vec{\imath} + mv_f \cos\theta\, \vec{\jmath}$$

Where i and j are unit vectors in the x and y directions respectively. Now the impulse is computed as follows,

$$\vec{J} = \vec{P}_f - \vec{P}_i == -mv_f \sin\theta\, \vec{\imath} + mv_f \cos\theta\, \vec{\jmath} - \left(-mv_f \sin\theta\, \vec{\imath} + mv_f \cos\theta\, \vec{\jmath}\right)$$

Since the final speed is equal to the initial speed, the x-components cancel each other. Thus, the impulse is along the y direction only, i.e.

$$\vec{J} = \Delta\vec{P} = 2mv\cos\theta\, \vec{\jmath} = 2(2.5kg)(10\, {}^{m}/_{s}) \cos 53\, \vec{\jmath} = \left(30\, {}^{kg\, m}/_{s}\right)\vec{\jmath}$$

(b) The force is obtained by using

$$\vec{F} = \frac{\vec{J}}{t} = \frac{(\Delta\vec{P})}{t} = \frac{\left(30\, {}^{kg\, m}/_{s}\right)\vec{\jmath}}{0.2s} = (150N)\vec{\jmath}$$

10. Below is a system known as a speed amplifier, where the two blocks are colliding elastically. Block 1 has a larger mass m_1=10kg, and block 2 a smaller mass m_2=0.5kg. Initially, block 1 is moving with velocity v_{1i}=5m/s and block 2 is stationary. Find the velocity of each particle after the collision.

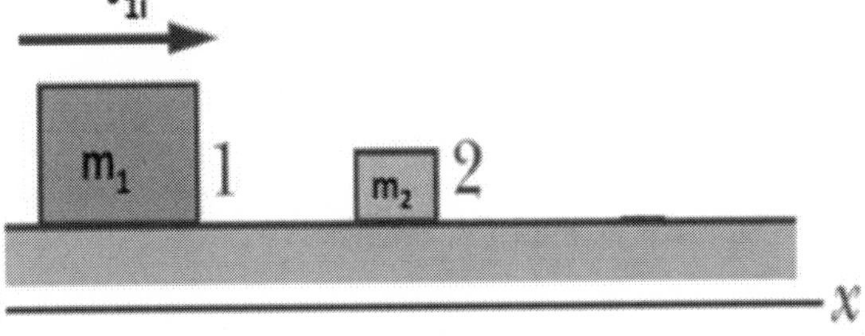

Solution:
Since there are no external forces, momentum is conserved. Because the collision is elastic, the kinetic energy is also conserved. Let the final velocities of the two masses after the collision be v_{1f} and v_{2f}. Applying conservation of momentum yields,

$$m_1 v_{1f} + m_2 v_{2f} = m_1 v_{1i} \qquad (1)$$

Solving for v_{2f} in terms of v_{1i} and v_{1f} yields

$$v_{2f} = \frac{m_1}{m_2}\left(v_{1i} - v_{1f}\right) \qquad (2)$$

Applying the conservation of kinetic energy results in

$$\frac{1}{2}m_1 v_{1f}^2 + \frac{1}{2}m_2 v_{2f}^2 = \frac{1}{2}m_1 v_{1i}^2 \qquad (3)$$

Substituting the value obtained for v_{2i} from Eq. (2) into Eq. (3) yields

$$v_{1f} = \frac{(m_1 - m_2)}{(m_1 + m_2)} v_{1i} = \frac{9.5kg}{10.5kg}(5\,{}^{m}/_{s}) = 4.52\,{}^{m}/_{s} \qquad (4)$$

Substituting the newly found value of v_{1f} into Eq. (2) solves for v_{2f}, i.e.

$$v_{2f} = \frac{m_1}{m_2}\left(v_{1i} - v_{1f}\right) = \frac{10kg}{0.5kg}(5\,{}^{m}/_{s} - 4.52\,{}^{m}/_{s}) = 9.52\,{}^{m}/_{s}$$

Additional Tidbits

i. How is the direction of the momentum of an object related to the direction of the motion of the object?
   a. It has the same direction as the velocity at all times.
   b. Sometimes it has opposite direction to the velocity.
   c. Sometimes it is perpendicular to direction of the velocity.
   d. Sometimes it is directed at a non-zero angle with the velocity.

ii. What kind of interactions are momentum conserving?
   a. Interactions that occur in the absence of external forces.
   b. Interactions that occur in the presence of frictional forces.
   c. Interactions that occur in the presence of the force of gravity.
   d. Interactions that occur in the presence of air resistance.

3) If a collision is elastic which of the following is true?
   a. Only momentum is conserved. b. Only kinetic energy is conserved.
   c. Both momentum and kinetic energy are conserved.
   d. The collision is taking place in the presence of friction.

4) In a momentum conserving process, what is the impulse?
(a) Twice the original. (b) The same as the original. (c) Zero. (d) Three time the original.

Answers To Tidbits

1. Momentum and velocity have the same direction at all times; hence choice (a) is the correct answer.
2. All interactions that take place in the absence of external forces are momentum-conserving interactions, hence choice (a) is the correct answer..
3. When a collision is elastic, it conserves momentum and kinetic energy.
4. Impulse represents the change in momentum, hence when the process is momentum conserving the impulse is zero, hence choice (c) is the correct answer.

Chapter Nine: Rotational Motion and Related Topics

1. What are the conditions for an object to be in complete equilibrium?
(a) The vector sum of all forces must be zero.
(b) The vector sum of all torques must be zero
(c) The vector sum of all the velocities is zero.
(d) Both (a) and (b).

Solution:
For point and extended objects there are two types of motion, i.e. translational and rotational. For an object to be in complete equilibrium both translational and rotational motion conditions must be ensured. While translational equilibrium is ensured when the vector sum of all forces is zero; the rotational equilibrium is achieved when the vector sum of all torques is zero. In other words for translational equilibrium the forces that are acting downwards balance all forces that are acting upwards; those pulling to the left are balanced by those pulling to the right.

For rotational motion to be ensured, all the torques that tend to rotate an object in a clockwise direction are balanced by those tending to rotate the object in the counter clock direction. Thus, choice (d) is the correct answer.

2. In a rotational equilibrium where should the axis of rotation be selected when calculating torques? (a) Only at the center of mass of the object. (b) Not at one end of the object (c) Not outside the object (d) Anywhere because the choice is arbitrary.

Solution:
In order to calculate torque, selecting an axis about which the object is rotating is needed. When the object is in equilibrium, the axis of rotation can be taken anywhere on the object or outside the object. Normally the choice is made about an axis that can facilitate the solution of the problem i.e., to eliminate as many torque terms as possible, which reduces the number of unknowns.

3. A wheel rotating with a constant angular acceleration turns through 30 revolutions during a 5s time interval. Its angular velocity at the end of this interval is 15 rad/s. What is the angular acceleration of the wheel?

Solution:
Fact #1: angular displacement is $\Delta\theta=30$ rev $=(30)(2\pi)$ rad$=60\ \pi$ rad,
Fact #2: time is 5s,
Fact #3: final angular velocity is $\omega=15$ rad
Fact #4: initial angular velocity is $\omega_0=$unknown,
Fact #5: angular acceleration is $\alpha=$unknown.

I have to list the known and unknown quantities and use the known facts to solve for the unknown. By definition the angular acceleration is the rate of change of angular velocity, i.e.

$$\alpha = \frac{\omega - \omega_0}{t} \qquad (1)$$

Here I have to take a detour because I have two unknowns and one equation. I have to obtain a second equation by using the information provided (as Prof. John would say by playing all my cards). I can use the angular displacement to find the average angular velocity, i.e.

$$\bar{\omega} = \frac{\Delta \theta}{\Delta t} = \frac{60\pi \text{ rad}}{5\text{s}} = 37.7 \, {}^{\text{rad}}\!/_{\text{s}} \qquad (2)$$

Since the angular acceleration is constant the average angular velocity is also given by

$$\bar{\omega} = \frac{\omega + \omega_0}{2} \qquad (3)$$

Substituting the value obtained in Eq. 2 into Eq. 3 yields

$$\omega + \omega_0 = 75.4 \, {}^{\text{rad}}\!/_{\text{s}} \qquad (4)$$

From Eq. 4, I can obtain the initial angular velocity, i.e.

$$\omega_0 = 75.4 \, {}^{\text{rad}}\!/_{\text{s}} - 15 \, {}^{\text{rad}}\!/_{\text{s}} = 60.4 \, {}^{\text{rad}}\!/_{\text{s}} \qquad (5)$$

This completes my detour. Now I go back to Eq. 1 and obtain the angular acceleration,

$$\alpha = \frac{\omega - \omega_0}{t} = \frac{15 \, {}^{\text{rad}}\!/_{\text{s}} - 60.4 \, {}^{\text{rad}}\!/_{\text{s}}}{5\text{s}} = -9.08 \, {}^{\text{rad}}\!/_{\text{s}^2}$$

4. Four rods each of length 10 m and mass 6kg are connected as shown in the figure below. Find the total moment of inertia of the system of rods about an axis passing through the end of the rod at O. Hint use of the moment of inertia of a rod of length L and mass M about an axis passing through its center, i.e.

$$I_{cm} = \frac{1}{12} ML^2$$

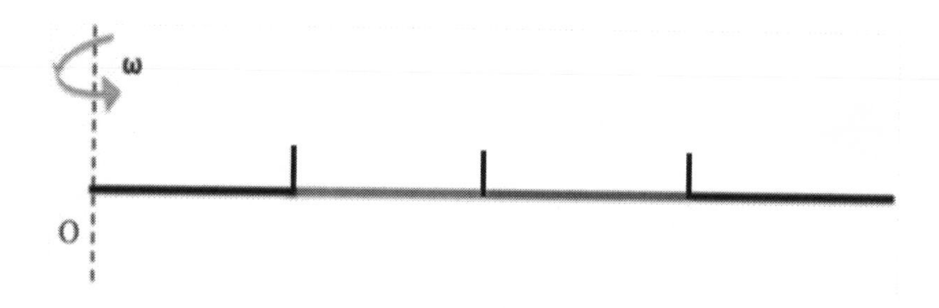

Solution:
Fact #1: mass of each rod is M=6kg,
Fact #2: length of each rod is L=10m,

The parallel axis theorem enables me to obtain the moment of inertia of an object about an axis passing through any point provided the moment inertia of the object about its center of mass is known. According to the parallel axis theorem; the moment of inertia of an object about an axis passing through a point outside other than the center of mass is given by

$$I_p = I_{cm} + Mh^2$$

Where h is the distance between the axis of rotation and the center of mass of the object. The fastest way to solve this problem is to combine all rods into a single rod having a total mass of four times the mass of each rod and a length 40m long. This means I can calculate the total moment of inertia in a single shot. Thus

$$I_p = \frac{1}{12}(24\text{kg})(40\text{m})^2 + (24\text{kg})(20\text{m})^2 = 12800\text{kgm}^2$$

The alternate approach would have been to calculate the moment of inertia of each and then obtain the total, which would have involved more work.

5. **The figure below shows an object (mass M=500kg) hanging by a rope (negligible mass) from a boom that consists of a uniform hinged beam (m=50kg) and horizontal cable (negligible mass) in complete equilibrium.**

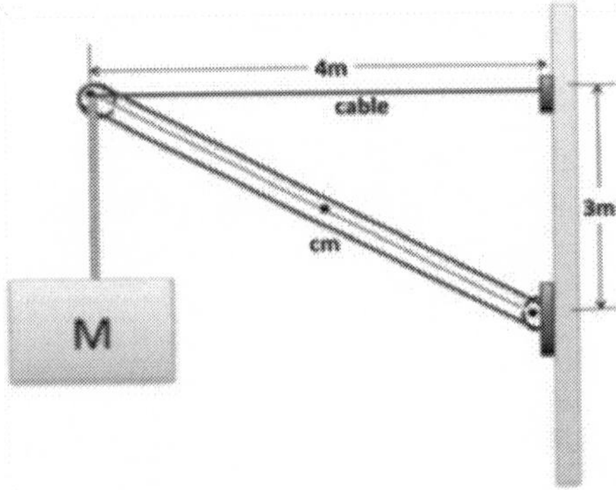

(a) List all forces acting on the beam. (b) By taking the axis of rotation at the hinge, find the total clockwise and counter clockwise torques separately. (c) Find the tension in the cable. (d) Find the magnitude of the force exerted by the wall.

Solution:
(a) The forces acting on the beam are:
The weight of the beam is mg (acting at the cm and downwards),
The weight of the object is Mg (downwards),
The tension in the cable is T (to the right).
The supporting force by the wall at the hinge is Fw.
The force exerted at the wall must have vertical and horizontal components, i.e. the vertical

component of the force exerted at the hinge prevents the beam from sliding down and the horizontal component acts to prevent the beam from piercing the wall.

(b) In order to calculate the clockwise and counter clockwise torques, I need to draw the free body diagram as indicated below.

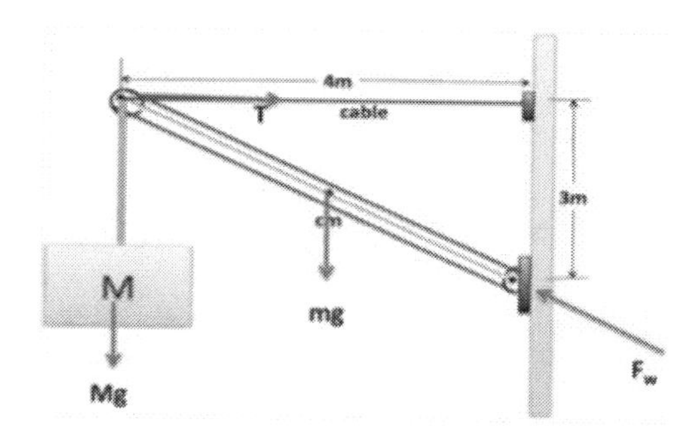

If the axis of rotation is fixed at the hinge, then F_w does not generate any torque. The weight of the object and the beam both generate a counter clockwise torque whereas the tension in the cable generates a clockwise torque. The magnitude of each torque is obtained by multiplying the force by the perpendicular distance to the axis of rotation. Thus the total clockwise torque is simply

$$\tau_{-} = T(3m) \qquad (1)$$

The total counterclockwise torque is given by

$$\tau_{+} = Mg(4m) + mg(2m) = 21000Nm \qquad (2)$$

Since the boom is in equilibrium, the clockwise and counterclockwise torques must be equal to each other. Thus equating Eqs. (1) and (2) and solving for the tension (T) yields,

$$T = 7000N \qquad (3)$$

(c) To find the force exerted by the wall, I have to apply the fact that the forces must be balanced. Since the force exerted by the wall is acting at an angle, it must be resolved into horizontal and vertical components.

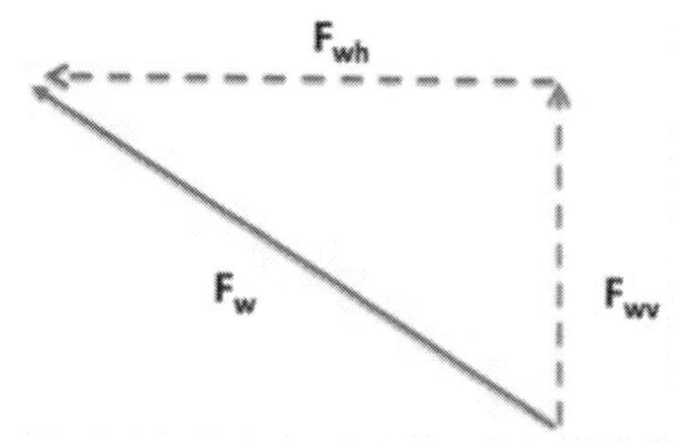

If I assign F_{wh} and F_{wv} to be the horizontal and vertical components of the force exerted by the wall as shown below, then, for the forces on the boom to be balanced the downward forces must be equal to the upward forces and the horizontal forces must be equal to each other, i.e.

$$F_{wv} = Mg + mg = (M + m)g = 5500N \qquad (4)$$

$$F_{wh} = T = 7000N \qquad (5)$$

Thus the magnitude of the force exerted on the wall is:

$$F_w = \sqrt{(F_{wh})^2 + (F_{wv})^2} = \sqrt{(7000N)^2 + (5500N)^2} = 8902N$$

6. The figure below shows a uniform hinged beam of length of 5m and a mass of 10kg held horizontally on one end with cable of negligible mass. If θ=30⁰, find (a) the tension in the cable and (b) the magnitude of the force on the hinge.

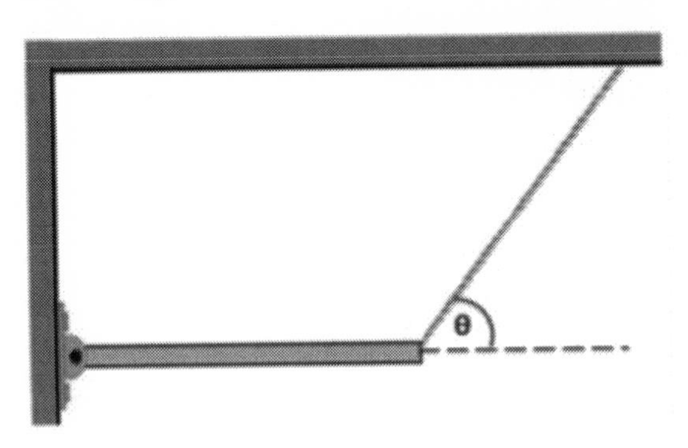

Solution:

Let me begin by identifying all the forces that are acting on the beam. These are the tension T in the cable, the weight of the beam acting downwards, and the force on the hinge. Unlike the sample problem the horizontal component of the force on hinge is directed towards the wall. These forces are acting as shown in the free body diagram below.

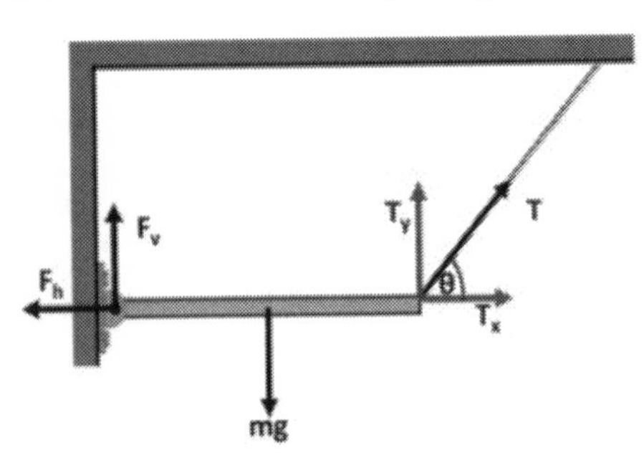

Writing the facts, I have:

Fact #1: mass, m=10kg,

Fact #2: Tension, T= unknown,

Fact #3: Force at the hinge, F= unknown.

Applying the first condition of the equilibrium yields,

$$\boldsymbol{T_x = F_h} \qquad (1)$$

$$\boldsymbol{T_y + F_v = mg} \qquad (2)$$

T_x, F_h, and F_v are unknowns. This means I need additional equations, which I can obtain them by applying the second condition of equilibrium. If I take the axis of rotation at the hinge then the weight of the beam generates a clockwise torque,

$$\tau_{cw} = mg\left(l/2\right) \qquad (3)$$

The vertical component of the tension (T_y) generates a counter clockwise torque,

$$\tau_{ccw} = T_y(l) \qquad (4)$$

Since the beam is in equilibrium, I can set Eqs. 3 and 4 equal to each other.

$$T_y(l) = mg\left(^{l}/_{2}\right) \qquad (5)$$

Solving for T_y yields,

$$T_y = \frac{mg}{2} = \frac{(10kg)(9.8\,^{m}/_{s^2})}{2} = 49\text{N} \qquad (6)$$

Substituting the value obtained for T_y in Eq. 2 yields,

$$F_v = mg - T_y = \frac{mg}{2} = 49N \qquad (7)$$

The vertical component of the tension (T_y) is related to the overall tension (T) by

$$T_y = T\sin\theta \Rightarrow T = \frac{T_y}{\sin\theta} = \frac{490N}{sin30} = 98N \qquad (8)$$

Since T_x is the horizontal component of the tension, then

$$T_x = T\cos\theta = 98N(\cos 30) = 84.87N \qquad (9)$$

Thus, the magnitude of the force on the hinge is

$$F = \sqrt{(T_x)^2 + (T_y)^2} = \sqrt{(84.87N)^2 + (49N)^2} = 98N \qquad (10)$$

7. A woman of mass 80 kg stands at the rim of a circular, frictionless platform of radius 5.0m that is rotating (without friction) with an angular speed of 100 rev/s. If the woman walks to the center of the platform and decreases the total moment of inertia of the system consisting of the platform and herself by 60%. (a) What is the basis of solving this problem, (b) what is the resulting angular speed of the platform, and (c) by what factor does the kinetic energy change?

Solution:

(a) Since there is no external torques that caused the change in the inertia, the basis of the solution to this problem is to apply conservation of angular momentum.

(b) Here I have the following facts;

Fact #1: initial angular velocity is 100 rev/s,

Fact #2: initial inertia is I_i=unknown,

Fact #3: final inertia is $I_f=0.4I_i$=unknown,

Fact #4: final angular velocity is unknown.

Even though there are more than one unknown, I will apply the conservation of momentum and see what happens. Let L_i and L_f represent the total angular momentum before and after respectively,

$$L_i = I_i\omega_i \qquad \text{and} \qquad L_f = I_f\omega_f$$

Since the angular momentum of the system is conserved, the initial and final angular momenta are equal to each other. In this way, I can solve for the final angular velocity,

$$\omega_f = \frac{I_i}{I_f}\omega_i = \frac{I_i}{0.4I_i}(100\ ^{rev}/_{s}) = 250\ ^{rev}/_{s}$$

(c) The initial and final rotational kinetic energies are given by

$$K_i = \frac{1}{2}I_i\omega_i^2 \qquad \text{and} \quad K_f = \frac{1}{2}I_f\omega_f^2$$

Now taking the ratio of the final kinetic energy to the initial yields,

$$\frac{K_f}{K_i} = \frac{\left(\frac{1}{2}I_f\omega_f^2\right)}{\left(\frac{1}{2}I_i\omega_i^2\right)} = 2.5$$

8. A wheel that has an initial angular velocity of 30 rad/s is slowing at a constant rate of 3rad/s^2 to a stop. By the time it stops, (a) how long has it been rotating, (b) through how many revolutions has it turned during that time?

Solution:
Let me list the facts of the problem.
Fact # 1: initial angular velocity is ω_0=30 dad/s,
Fact # 2: final angular velocity is 0,
Fact # 3: angular acceleration is -3 rad/s^2,
Fact # 4: duration of rotation is t=unknown,
Fact #5: angular displacement is $\Delta\theta$=unknown.

The time duration is calculated from the definition of acceleration, i.e.

$$\alpha = \frac{\omega - \omega_0}{t} \Rightarrow t = \frac{\omega - \omega_0}{\alpha} = \frac{0 - 30\ ^{rad}/_{s}}{-3\ ^{rad}/_{s^2}} = 10s$$

(b) There are several options to find the angular displacement. The simplest approach is to use the average angular velocity.

$$\Delta\theta = \bar{\omega}t = \left(\frac{\omega + \omega_0}{2}\right)t = \left(15\ ^{rad}/_{s}\right)(10s) = 150\ \text{rad}$$

To change the angular displacement into revolutions, I have to simply divide by 2π radians. Thus

$$\Delta\theta = \frac{\mathbf{150\ rad}}{\mathbf{2\pi\ ^{rad}/_{rev}}} \simeq 23.87\ \text{revs}$$

9. A uniform 25kg thin rod of length 12 m is standing vertically as shown below. If the rod is let go and allowed to fall, determine the angular speed with which it hits the ground. (Hint: assume the bottom of the rod is fixed and the inertia of a rod about its center is I_{cm}=(1/12)ML^2)

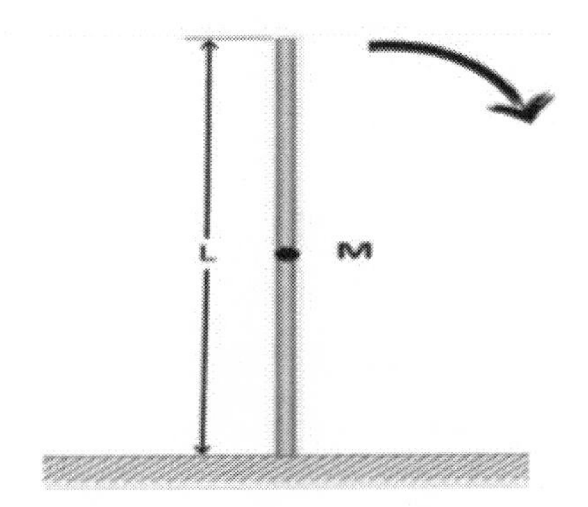

Solution:
When the rod is completely on the ground, its center of mass has fallen half of the length of the rod. This problem is easily solved using energy concepts. Taking the bottom of the rod as a reference point, the initial energy is all gravitational potential energy and it is given by

$$E_i = U_i = Mgh = Mg\left(\frac{L}{2}\right) \qquad (1)$$

When the rod hits the ground, all its gravitational potential energy is converted to rotational kinetic energy, i.e.

$$E_f = K_f = \frac{1}{2}I\omega^2 \qquad (2)$$

Since the rod does not slip, the rotation is taking place about an axis passing through its bottom. The moment of inertia of the rod about an axis passing through its bottom can be calculated using the parallel axis theorem, i.e.

$$I = I_{cm} + Mh^2 = \frac{1}{12}ML^2 + M\left(\frac{L}{2}\right)^2 = \frac{1}{3}ML^2 \qquad (3)$$

Substituting the formula obtained in Eq. 3, into Eq. 2 yields

$$K_f = \frac{1}{2}\left(\frac{1}{3}ML^2\right)\omega^2 = \frac{1}{6}ML^2\omega^2 \qquad (4)$$

Since there is no energy loss, I can equate Eqs. 4 and 1 to solve for the angular velocity.

$$\frac{1}{6}ML^2\omega^2 = Mg\left(\frac{L}{2}\right) \quad \Rightarrow \omega = \sqrt{\frac{3g}{L}} = 1.58\,{}^{rad}/{}_{s}$$

10. The figure below shows a uniform disk with mass M = 6kg and radius R=10cm mounted on a fixed horizontal axle. A block with mass m = 2 kg hangs from a massless cord that is wrapped around the rim of the disk. Find (a) the acceleration of the falling block, (b) the angular acceleration of the disk, and (c) the tension in the cord. (Hint: use inertia of a disk =(1/2)MR^2)

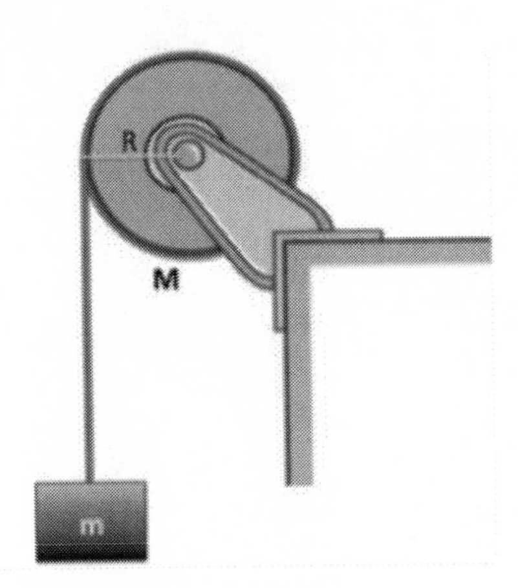

Solution:
(a) Before I proceed with any calculations, I have to identify the forces acting on each object and I have to remember that while the block is describing translational motion, the disk is undergoing rotational motion. The forces acting on the block are the tension in the cord and the weight of the block. The free body diagram of the block is drawn as shown below.

Similarly, the free body diagram of the disk is,

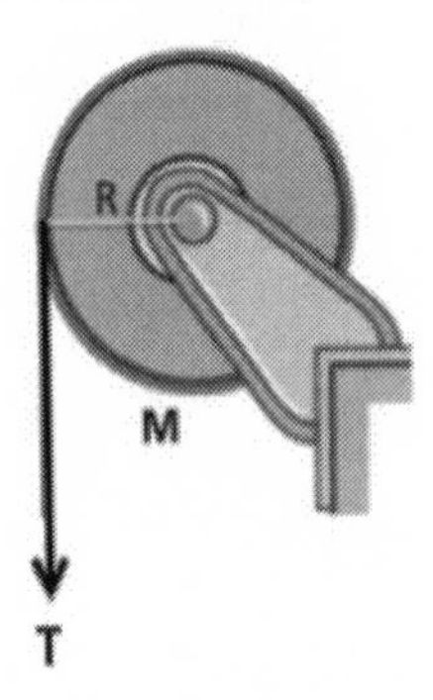

As shown on the free body diagram, there are two forces acting on the block and there is a single force on the disk. The tension is acting on the rim of the disk, and it is causing the disk to rotate in a counterclockwise direction. Now I have to find the net force on each object. Taking the direction of motion as positive, the net force on the block is

$$F = mg - T = ma \qquad (1)$$

The torque on the disk is the product of the force (T) and the lever arm, which is the radius of the disk, i.e.

$$\tau = TR = I\alpha \qquad (2)$$

Where α is the angular acceleration of the disk, which is related to the translational acceleration by the following.

$$\alpha = \frac{a}{R} \qquad (3)$$

Rewriting Eq. 2 and replacing I by the moment of inertia of the disk gives

$$T = \frac{1}{2}Ma \qquad (4)$$

Substituting this expression into Eq. 1 and solving for the acceleration yields

$$a = \frac{2mg}{M + 2m} = \frac{2(2kg)(10\,{}^{m}/_{s^2})}{10kg} = 4\,{}^{m}/_{s^2}$$

(b) Here I have to use Eq. 3 to find the angular acceleration, i.e.

$$\alpha = \frac{a}{R} = \frac{4\,{}^{m}/_{s^2}}{0.1m} = 40\,{}^{rad}/_{s^2}$$

(c) Finally the tension T, is found by using Eqs. 4,

$$T = \frac{1}{2}Ma = (0.5)(6kg)\left(4\,{}^{m}/_{s^2}\right) = 12.0N$$

11. A small, solid sphere of mass 2.0 kg and radius 1.0 m rolls without slipping along a track consisting of a slope and a loop-the-loop with radius R=4m. It starts from rest near the top of the track at a height of h, where h is large compared to the radius of the sphere. What is the minimum value of h (in terms of the radius of the loop R) such that the sphere completes the loop? (Hint: use the moment of inertia of a sphere I=(2/5)MR²).

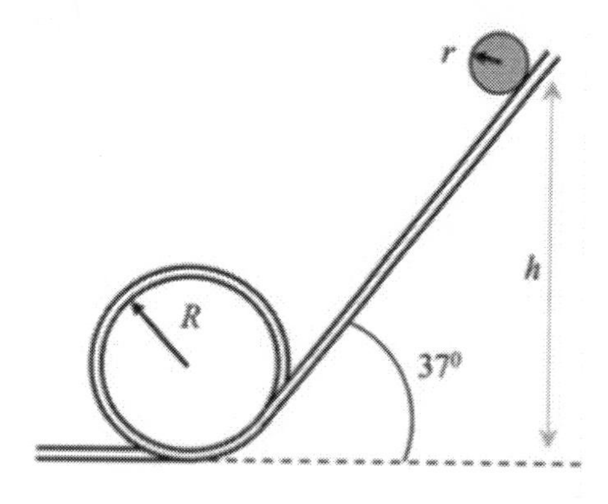

Solution:
This problem is easily handled using energy considerations. Assuming the sphere to be uniform, its center of mass must be located at its geometric center. This means the mass of the sphere is a distance of H=h+r from the bottom of the loop. By taking the reference of the potential energy at the bottom of the loop to be zero, then the initial mechanical energy is,

$$E_i = mgH = mg(h + r) \qquad (1)$$

When the sphere is at the top of the loop-the-loop, the energy is partly gravitational, partly translational kinetic and partly rotational kinetic energy. Hence

$$E_f = mg(2R - r) + \frac{1}{2}mv^2 + \frac{1}{2}I\omega^2 \qquad (2)$$

Making the following substitutions,

$$I = \frac{2}{5}mr^2, \qquad \omega = \frac{v}{r}$$

and simplifying yields,

$$v^2 = \frac{10gh + 20gr - 20gR}{7} \qquad (3)$$

When v is the minimum speed of the sphere required to complete the loop, there is no normal reaction force at the top of the loop. This means the weight of the sphere is the only centripetal force. Applying newton's law when the sphere is at the loop-the-loop yields

$$m\frac{v^2}{R - r} = mg \;\Rightarrow v^2 = g(R - r) \qquad (4)$$

Equating Eqs. 3 and 4 and solving for the height yields,

$$\frac{10gh + 20gr - 20gR}{7} = g(R - r)$$

Hence

$$h = \frac{27R - 27r}{10} = \frac{27(4m) - 27(1m)}{10} = 8.1m$$

12. A merry-go-round of radius 5.0 m has a moment of inertia 100kgm² and is rotating at 120 rev/min about a frictionless vertical axle. A 40 kg child hops onto the merry-go-round and manages to sit down on its rim. What is the new angular velocity of the merry-go-round?

Solution:
Now listing the facts of the problems I have:
Fact #1: radius of the sphere is 5m,
Fact #2: initial angular velocity is ω_0= 120 rev/min,
Fact #3: initial moment of inertia is I_0=100kgm^2,
Fact #4: final moment of inertia is I_f= unknown
Fact #5: Final angular velocity is ω= unknown.

Before I can apply conservation of momentum, I have to find the combined inertia of the child and the merry-go-round. Assuming the child to be a point object, his moment of inertia is obtained using

$$I_b = m_b r^2 = 20kg(5m)^2 = 500kgm^2$$

The new inertia of the child-merry-go-round system is thus

$$I_f = 100kgm^2 + 500kgm^2 = 600kgm^2$$

According to the conservation of momentum I can write

$$L_f = L_i \Rightarrow I_f\omega_f = I_0\omega_0$$

Solving for the final angular velocity yields,

$$\omega_f = \frac{(100\text{kgm}^2)(120\,{}^{\text{rev}}/_{\text{min}})}{(600\text{kgm}^2)} = 20\,{}^{\text{rev}}/_{\text{min}}$$

Additional Tidbits

1) What is moment of inertia?
   (a) It is the degree of resistance of an object to transitional motion.
   (b) It is the degree of resistance of an object to rotational motion.
   (c) It is the degree of resistance of an object to both transitional and rotational motion.
   (d) None of the above.

2) What are the factors that affect the moment of inertia of an object?
(a) Its shape. (b) Its mass. (c) The distance of the object to the axis of rotation.
(d) All of the above.

3) What is torque?
(a) It is the power to cause rotational motion. (b) It is the power to cause translational motion.
(c) It is the momentum of the object. (d) It is the net force on an object.

4) What are the factors that affect the torque of an object?
   (a) The magnitude of the force,
   (b) The distance between the axis of rotation and the location of the force.
   (c) The angle between the direction of the force and the vector from the axis of rotation to where the force is applied. (d) All of the above

5) Does a force generate torque at all times? Explain why or why not?

6) Describe why an object moving on a circular path at constant speed must have a non-zero acceleration and a non-constant velocity. Specify the directions of the acceleration and velocity.

7) What is the moment of inertia of a rod of negligible mass?
   (a) It is zero. (b) It cannot be determined.
   (c) It depends on the axis of rotation. (d) It has inertia only about an axis passing through the center of mass.

8) If the moment of inertia of a meter stick of mass 120kg about an axis passing through its center of mass is 10kgm^2, what is its inertia about an axis passing through one of its ends?
a. 20kgm^2 b. 30kgm^2 c. 40kgm^2 d. 50kgm^2

9) In an angular momentum conserving interaction, what happens to the angular velocity if the momentum of inertia is decreased?

(a) It remains the same.
(b) It also decreases by the same amount.
(c) It increases.
(d) It decreases by a factor of 2.

Answers To Tidbits

(a)The moment of inertia of an object is a measure of the degree of resistance of an object to rotational motion; hence choice (b) is the correct answer.

(b) The moment of inertia of an object depends on the mass, shape and the distance of the object to the axis of rotation; hence choice (d) is the correct answer.

(c) Torque is the power of force to cause rotational motion.

(d) A force does not always generate torque. A force can only generate torque if it has a non-zero lever arm.

(e) The magnitude of the torque depends on the force, the lever arm and the direction of the force acting on the object.

(f) Even though the speed is constant; the direction of motion is continuously changing. As a result the change of velocity is not zero. This causes the object to have a nonzero acceleration; the acceleration is always directed towards the center.

(g) Since an object must have a mass in order to have moment of inertia; the moment of inertia of a mass less object is zero.

(h) The moment of inertia of an object that is rotating about an off center point is obtained by using the parallel axis theorem, i.e.

$$I_p = I_{cm} + mh^2 \Rightarrow I_p = 10\text{kg}m^2 + (120kg)(0.5m)^2 = 40\text{kg}m^2$$

(i) If the momentum of inertia of a rotating object is decreased then its angular velocity is gets increased, hence choice (c) is the correct answer.

Chapter Ten: Gravitation Force

1. How much will an astronaut who has a weight of 900N weigh on the moon where the acceleration due to gravity is approximately one-sixth that of the Earth?

Solution:

Fact #1: acceleration due to gravity of Earth=$g \simeq 10 m/s^2$,
Fact #2: weight on Earth is $W==mg=900N$,
Fact #3: acceleration due to gravity is $g_m=(1/6)g$.
Fact #4: weight on moon is w=unknown.
The weight of an object at any location or on any planet is the product of its mass and the acceleration due to gravity at that location. Thus the weight of the astronaut on the moon is calculated as follows

$$w = mg_m = m(^1/_6)g = \frac{mg}{6} = \frac{900N}{6} = 150N$$

Even though his weight is much smaller than when he is on the earth, the mass of the astronaut remains the same.

2. If the distance between two objects of masses m_1 and m_2 is tripled by what factor does the gravitational attraction between the two objects change?

Solution:
Fact #1: initial separation is r,
Fact #2: new separation is $r_{new}=3r$,
Fact #3: new force is F_{new}=unknown,
Newton's law of gravitation gives the magnitude of the gravitational force between any two objects of mass m_1 and m_2 separated by a distance r, i.e.

$$F = G\frac{m_1 m_2}{r^2} \qquad (1)$$

Applying Newton's law of gravitation to the new situation yields,

$$F_{new} = G\frac{m_1 m_2}{(r_{new})^2} = G\frac{m_1 m_2}{(3r)^2} = \frac{1}{9}G\frac{m_1 m_2}{r^2} = \frac{1}{9}F \qquad (2)$$

Thus, the force will be reduced by a factor of 9.

3. Five objects each of mass 5 kg are located at points in the xy plane as shown in the figure below. Find the magnitude and direction of the net gravitational force on the object at the origin. (use: $G= 6.6726 \times 10^{-11}\ N \cdot m^2/kg^2$).

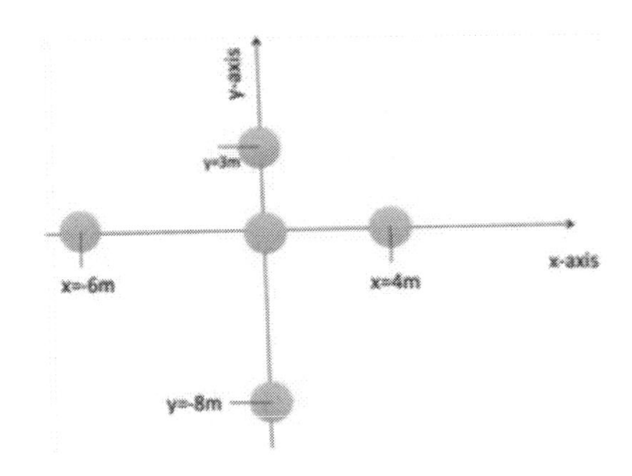

Solution:
Fact #1: universal gravitational constant is $G=6.672x10^{-11}N.m^2/kg^2$,
Fact #2: distance of each object given
Fact #3: mass of each object is given.
According to Newton's law of gravitation, each object exerts a gravitational pull on the other objects independent of the presence of other objects. The magnitude of each force is given by

$$F = G\frac{m_i m_j}{r_{ij}^2} \qquad (1)$$

where i and j represent the ith and j^{th} object and r_{ij} is the distance between the centers of each object. Let F_1, F_2, F_3, and F_4 be the forces due to the masses located at (4m, 0m), (0m, 3m), (-6m, 0m), and (0m, -8m) respectively. Applying Newton's law gives the four forces,

$$F_1 = 6.6726 \times 10^{-11}\ N\ {}^{m^2}/_{kg^2}\left(\frac{(5kg)(5kg)}{(4m)^2}\right) = 1.04 \times 10^{-10}N$$

$$F_2 = 6.6726 \times 10^{-11}\ N\ {}^{m^2}/_{kg^2}\left(\frac{(5kg)(5kg)}{(3m)^2}\right) = 1.86 \times 10^{-10}N$$

$$F_3 = 6.6726 \times 10^{-11}\ N\ {}^{m^2}/_{kg^2}\left(\frac{(5kg)(5kg)}{(-6m)^2}\right) = 4.63 \times 10^{-11}N$$

$$F_4 = 6.6726 \times 10^{-11}\ N\ {}^{m^2}/_{kg^2}\left(\frac{(5kg)(5kg)}{(-8m)^2}\right) = 2.61 \times 10^{-11}N$$

Since the gravitational force is always attractive, each force is directed towards the pulling object. This means F_1 and F_3 are opposite to each other and F_2 and F_4 are also opposite to each other. Thus the net force along the x and y direction are given by

$$F_x = F_1 - F_3 = 5.77 \times 10^{-11}N$$

$$F_y = F_2 - F_4 = 15.93 \times 10^{-11}N$$

The magnitude of the net force is obtained by Pythagoras theorem,

$$F_{net} = \sqrt{(5.77 \times 10^{-11}N)^2 + (15.93 \times 10^{-11}N)^2} = 5.36 \times 10^{-10}N$$

The direction of the net force is obtained by taking the inverse tangent of the ratio of the y-component to that of the x-component, i.e.

$$\theta = \tan^{-1}\frac{F_y}{Fx} = \frac{15.93}{5.77} \Rightarrow \theta \simeq 70^0$$

4. An object of mass m is dropped from rest from a height 5x10^6 m above the surface of the earth. If the object falls in the absence of air resistance, what is its speed when it strikes the earth?

Solution
Fact #1: distance of object above the surface is h=5x10^{6}m,
Fact #2: radius of Earth is R=6.37x10^{6}m,
Fact #3: mass of the Earth is M=5.972x10^{24} kg,
Fact #4: universal gravitational constant is G=6.672x10^{-11}N.m^2/kg^2.

Applying the conservation of mechanical energy easily solves this problem. Since the object is falling under the influence of gravity due the Earth, the initial gravitational energy is given by

$$U_i = -G\frac{Mm}{r} \qquad (1)$$

where M is the mass of the Earth and r is the initial distance between the center of the Earth and the object,

$$r = h + R = 11.37 \times 10^6 m$$

Since the object was dropped, its initial velocity is zero. This means the corresponding initial kinetic energy is zero, i.e.

$$K_i = 0 \qquad (2)$$

The total initial mechanical energy is thus

$$E_i = K_i + U_i = -G\frac{Mm}{r} \qquad (3)$$

When the object strikes the surface of the Earth, it has speed, hence the final mechanical energy is obtained by

$$E_f = K_f + U_f \Rightarrow E_f = \frac{1}{2}mv^2 - G\frac{Mm}{R} \qquad (4)$$

Here R is the radius of the Earth. Now applying conservation of mechanical energy yields

$$\frac{1}{2}mv^2 - G\frac{Mm}{R} = -G\frac{Mm}{r} \qquad (5)$$

Solving for the speed yields

$$v = \sqrt{-2G\frac{M}{r} + 2G\frac{M}{R} =} = \sqrt{2GM\left(\frac{1}{R} - \frac{1}{r}\right)} \simeq 7.4 \times 10^3\ ^{m}/_{s}.$$

5. If the planet Saturn has a mass 95.2 times that of the earth and a radius 9.47 times that of the earth, then what will be the escape velocity of an object so that it can go beyond the influence of the gravitational force of Saturn?

Solution:
Let me begin by listing the facts of the problem.
Fact #1: mass of Earth is $M=5.972x10^{24}$kg,
Fact #1: mass of Saturn is $M_S=95.2M$,
Fact #3: radius of Earth is $R= 6.37x10^6$m,
Fact #4: radius of Saturn is $R_S=9.47R$.
Fact #5: universal gravitational constant is $G=6.672x10^{-11}N.m^2/kg^2$,
Fact #6: escape velocity is v_0=unknown,
Fact #7: mass of object is m=unknown.

As before, I will be able to solve this problem by applying the conservation of mechanical energy. The initial mechanical energy is the sum of the initial kinetic and gravitation potential energies, i.e.

$$E_i = \frac{1}{2}mv_0^2 - G\frac{M_S m}{r_S} \qquad (1)$$

The final mechanical energy is also the sum of the kinetic and gravitational potential energies after escaping Saturn's gravitational influence, i.e.

$$E_f = \frac{1}{2}mv_f^2 - G\frac{M_S m}{r} \qquad (2)$$

Equating Eqs. 1 and 2 yields,

$$\frac{1}{2}mv_0^2 - G\frac{M_S m}{r_S} = \frac{1}{2}mv_f^2 - G\frac{M_S m}{r} \qquad (3)$$

If v_0 is the escape velocity, then

$$v_f = 0 \quad \text{and} \quad r = \infty \qquad (4)$$

Substituting these values into Eq. 3 and solving for the escape velocity yields,

$$v_0 = \sqrt{2G\frac{M_s}{r_s}} = \sqrt{2G\frac{95.2M}{9.47R}} = \sqrt{2(6.672 \times 10^{-11}N.\frac{m^2}{kg^2})\frac{(95.2)5.972 \times 10^{24}kg}{9.47 \times 6.37 \times 10^6 m}}$$
$$= 35.463\ ^{km}/_{s}$$

6. A newly discovered asteroid is rotating around the Sun in a nearly circular orbit at an average distance of 6 times the Earth-Sun distance. If the mass of the asteroid is five times that of the Earth, then what is the period of the asteroid?

Solution:

Fact #1: Earth-Sun distance is $R_E \simeq 1.5 \times 10^{11}$m

Fact #2: asteroid distance from the Sun is $R_A = 6R_E \simeq 9 \times 10^{11}$m,
Fact #3: mass of the Earth is $M = 5.972 \times 10^{24}$kg,
Fact #4: mass of asteroid is $5M = 29.86 \times 10^{24}$kg,
Fact #5: period of the Earth is $T_E = 365.25$ days,
Fact # 6: period of asteroid is $T_A =$ unknown,

This problem is solved by Kepler's harmonic law, which states that the ratio of square of the period of a planet around the Sun to the cube of the mean distance from the sun is constant, i.e.

$$\frac{T^2}{R^3} = \text{constant} \qquad (1)$$

According to Eq. 1, the period can be calculated if the constant is known. However, the easier approach is to relate the period of the asteroid to that of the Earth. Thus Eq. 1 can be written equivalently as follows,

$$\frac{T_A^2}{R_A^3} = \frac{T_E^2}{R^3} \qquad (2)$$

Solving for the period of the asteroid yields,

$$T_A = \sqrt{\frac{R_A^3}{R^3}}\, T_E = \sqrt{\left(\frac{9 \times 10^{11}\text{m}}{1.5 \times 10^{11}\text{m}}\right)^3}\ 365.25 \text{ days} \simeq 5368 \text{ days} = 14.7 \text{ years}$$

7. If the distance between a planet and the sun is doubled by what factor does the period of the planet change?

Solution:
Fact #1: initial distance is R,
Fact #2: new distance $R_{new} = 2R$.

As in Problem 6, Kepler's third law gives the period of a planet around the sun, i.e.

$$T^2 = (\text{constant})R^3 \qquad (1)$$

As the distance between the sun and the planet changes, its period also changes. Thus the new period is given by,

$$T_{new}^2 = (\text{constant})(R_{new})^3 \qquad (2)$$

Taking the ratio of Eq. 2 to Eq. 1 yields

$$\frac{T_{new}^2}{T^2} = \frac{(\text{constant})(R_{new})^3}{(\text{constant})R^3} = \left(\frac{R_{new}}{R}\right)^3 = \left(\frac{2R}{R}\right)^3 = 8$$

Therefore, the new period is

$$\frac{T_{new}}{T} = \sqrt{8} = 2.83$$

Thus, the period increases by a factor 2.83.

8. A 300 kg geostationary satellite is orbiting in space on a circular orbit at an altitude of 36,000km above the surface of the Earth. (a) Find the kinetic energy of the satellite, (b) find the period of the satellite, (c) how does the satellite appear as seen by a ground-based observer and why does it appear that way, and (d) find the speed of the satellite.

Solution:
Fact #1: mass of satellite is m=300kg,
Fact #2: altitude of satellite is h=36,000km=$3.6\text{x}10^7$m,
Fact #3: mass of the Earth is M=$5.972\text{x}10^{24}$ kg,
Fact #4: radius of Earth is R=$6.37\text{x}10^6$m,
Fact #5: universal gravitational constant is G=$6.672\text{x}10^{-11}\text{N.m}^2/\text{kg}^2$,
Fact #6: distance of satellite from the center of the Earth r = R+h = $4.237\text{x}10^7$m,
Fact #7: kinetic energy K=unknown,
Fact #8: period of satellite = unknown.

(a) The satellite is orbiting the earth due to the gravitational force between the two objects. The magnitude of the gravitational force is given by

$$F = G\frac{Mm}{r^2} \qquad (1)$$

According to Newton's second law of motion, the force is given by

$$F = ma \qquad (2)$$

The acceleration of an object undergoing circular motion is

$$a = \frac{v^2}{r} \qquad (3)$$

Substituting for the acceleration from Eq. 3 into Eq. 2 yields,

$$F = m\frac{v^2}{r} \qquad (4)$$

Equating Eqs. 4 and 1 and solving for the speed yields,

$$v^2 = \frac{GM}{r} \qquad (5)$$

Thus, the kinetic energy is

$$K = \frac{1}{2}mv^2 = \frac{1}{2}m\frac{GM}{r} = \frac{(0.5)(300\text{kg})\left(6.672\times10^{-11}\,\text{N.m}^2/\text{kg}^2\right)(5.972\times10^{24}\text{kg})}{4.237\times10^7\text{m}} = 1.41\times10^9\text{J}$$

(b) The period is the circumference of the orbit divided by the speed of the satellite, i.e.

$$T = \frac{2\pi r}{v} = \frac{2\pi r}{\sqrt{\frac{GM}{r}}} = 2\pi\sqrt{\frac{r^3}{GM}} = 2\pi\sqrt{\frac{(4.237\times10^7\text{m})^3}{\left(6.672\times10^{-11}\,\text{N.m}^2/\text{kg}^2\right)(5.972\times10^{24}\text{kg})}} = 86{,}812\text{ s} \cong 24\text{ hours}$$

(c) Since the satellite's period the same as the time it takes for Earth to rotate on its axis, the satellite appears to be stationary.

(d) The speed of the satellite is given by

$$v = \sqrt{\frac{GM}{r}} = \frac{2\pi r}{T} = \frac{(2\pi)4.237 \times 10^7 m}{86{,}812s} = 3066.6 \, {}^{m}/_{s}$$

9. A spacecraft of mass 10,000kg, which had landed on the moon's surface, is preparing to leave the moon so that it can return to earth. If the mass and radius of the moon are $7.35x10^{22}$ kg and $1.77x10^3$ km respectively, how fast should the spacecraft blast off the surface of the moon so that it can escape the Moon's gravitational field?

Solution:
Fact #1: mass of space craft=m=10, 0000kg,
Fact #2: mass of the moon=M=$7.35x10^{22}$ kg,
Fact #3: radius of the moon=R=$1.77x10^3$ km=$1.77x10^6$m,
Fact #4: initial speed of spacecraft =v_0=unknown.

This question is similar to problem 5, which means I will apply the same principles to find the escape velocity of the spacecraft from the moon. The initial mechanical energy of the spacecraft is the sum of its kinetic and gravitational energies, i.e.

$$E_i = \frac{1}{2}mv_0^2 - G\frac{Mm}{R}$$

The final energy will also be the sum of the final kinetic and gravitational potential energies, i.e.

$$E_f = \frac{1}{2}mv^2 - G\frac{Mm}{r}$$

where v and r are the speed of the spacecraft and its distance from the center of the moon as it leaves the gravitational field. If v_0 is the escape velocity, then the final velocity is zero. When the spacecraft leaves the gravitational field, its distance from the center of the moon is much greater than the radius of the moon, i.e.

$$r \ggg R$$

So that

$$\frac{1}{r} \approxeq 0$$

Thus the final energy of the spacecraft is zero. Now applying conservation of energy leads to

$$\frac{1}{2}mv_0^2 - G\frac{Mm}{R} = 0 \qquad (1)$$

Solving for the escape speed yields,

$$v_0 = \sqrt{2G\frac{M}{R}} = \sqrt{2 \times 6.672 \times 10^{-11} \, N.m^2/_{kg^2} \frac{7.35x10^{22} \, kg}{1.77 \times 10^6 m}} \approxeq 2.35 \, {}^{km}/_{s}$$

10. A manmade satellite of mass 500kg is revolving in a circular orbit at an altitude of 300km above the earth's surface. How much work would be required to take the satellite to a new orbit whose altitude is twice the current distance above the surface of the earth?

Solution:

Fact #1: mass of satellite is m=500kg,

Fact #2: mass of the earth is $M=5.972\text{x}10^{24}$ kg,

Fact #3: radius of the earth is $R=6.37\text{x}10^{6}$m

Fact #4: initial earth-satellite distance is $r_1=300{,}000\text{m}+R=6.37\text{x}10^{6}$m,

Fact #5: new earth-satellite distance is $r_2=6.97\text{x}10^{6}$m.

I know that the work done on the satellite is the same as the sum of the changes in its kinetic and gravitational energies, i.e.

$$W = \Delta K + \Delta U \qquad (1)$$

This means I have to find the initial and final kinetic energies in order to find the difference. The total mechanical energy of a satellite revolving along a circular orbit at a distance r is the sum of its kinetic and gravitational potential energies, i.e.

$$E = \frac{1}{2}mv^2 - G\frac{Mm}{r} \qquad (2)$$

The centripetal force responsible for making the satellite rotate around the earth is the gravitational force between the satellite and the Earth, i.e.

$$F = G\frac{Mm}{r^2} \qquad (3)$$

According to Newton's second law of motion, the net force is given by

$$F = ma = m\frac{v^2}{r} \qquad (4)$$

Equating Eqs. 4 and 3, and solving for v yields,

$$m\frac{v^2}{r} = G\frac{Mm}{r^2} \Rightarrow \quad v^2 = G\frac{M}{r} \qquad (5)$$

By substituting the results obtained in Eq. 5 into Eq. 2 results in

$$E = \frac{1}{2}mv^2 - G\frac{Mm}{r} \quad \Rightarrow E = \frac{1}{2}m\left(G\frac{M}{r}\right) - G\frac{Mm}{r} = -G\frac{Mm}{2r}$$

The last equation gives the mechanical energy in terms of the masses of the earth and the satellite, and the distance between the center of the earth and the satellite. Thus the two energies can be written as

$$E_i = -G\frac{Mm}{2r_1}$$

$$E_f = -G\frac{Mm}{2r_2}$$

Now calculating the difference between the two orbits gives

$$E_f - E_i == -G\frac{Mm}{2r_2} + G\frac{Mm}{2r_1} = \frac{1}{2}GMm\left(\frac{1}{r_1} - \frac{1}{r_2}\right)$$

Upon substituting the values for each variable yields,

$$E_f - E_i = (0.5)\left(6.672 \times 10^{-11}\, N.m^2/_{kg^2}\right)(5.972x10^{24}\ kg)\left(\frac{1}{10^6 m}\right)\left(\frac{1}{6.67} - \frac{1}{6.97}\right) = 1.29 \times 10^7 J$$

Additional Tidbits

1. What is the nature of the gravitational force between two objects?
   (a) It is directly proportional to the masses of the objects.
   (b) It is inversely proportional to the square of the distance between the two objects.
   (c) It is always attractive.
   (d) It is always repulsive
   (e) All of the above except (d).

Answers to Tidbits

1. The gravitational force between two objects is always attractive, depends directly on the masses of the attracting objects and it is proportional to the inverse square of the distance between the centers of the two objects.